GETTING TO KNOW
Mobile GIS

Pinde Fu

Esri Press
REDLANDS | CALIFORNIA

Esri Press, 380 New York Street, Redlands, California 92373-8100.
Copyright © 2025 Esri.
All rights reserved.
Printed in the United States of America.
29 28 27 26 25   1 2 3 4 5 6 7 8 9 10

ISBN: 9781589488076
Library of Congress Control Number: 2024945555

The information contained in this document is the exclusive property of Esri or its licensors. This work is protected under United States copyright law and other international copyright treaties and conventions. No part of this work may be reproduced or transmitted in any form or by any means, electronic or mechanical, including photocopying and recording, or by any information storage or retrieval system, except as expressly permitted in writing by Esri. All requests should be sent to Attention: Director, Contracts and Legal Department, Esri, 380 New York Street, Redlands, California 92373-8100, USA.

The information contained in this document is subject to change without notice.

US Government Restricted/Limited Rights: Any software, documentation, and/or data delivered hereunder is subject to the terms of the License Agreement. The commercial license rights in the License Agreement strictly govern Licensee's use, reproduction, or disclosure of the software, data, and documentation. In no event shall the US Government acquire greater than RESTRICTED/LIMITED RIGHTS. At a minimum, use, duplication, or disclosure by the US Government is subject to restrictions as set forth in FAR §52.227-14 Alternates I, II, and III (DEC 2007); FAR §52.227-19(b) (DEC 2007) and/or FAR §12.211/12.212 (Commercial Technical Data/Computer Software); and DFARS §252.227-7015 (DEC 2011) (Technical Data – Commercial Items) and/or DFARS §227.7202 (Commercial Computer Software and Commercial Computer Software Documentation), as applicable. Contractor/Manufacturer is Esri, 380 New York Street, Redlands, California 92373-8100, USA.

Esri products or services referenced in this publication are trademarks, service marks, or registered marks of Esri in the United States, the European Community, or certain other jurisdictions. To learn more about Esri marks, go to: links.esri.com/EsriProductNamingGuide. other companies and products or services mentioned herein may be trademarks, service marks, or registered marks of their respective mark owners.

For purchasing and distribution options (both domestic and international), please visit esripress.esri.com.

Contents

Preface .... *xi*
Acknowledgments .... *xiii*
How to use this book .... *xv*

Chapter 1: Mobile GIS overview and mapcentric data collection using ArcGIS Field Maps . . 1
    Objectives 1
    Introduction 1
    Mobile GIS: Concepts and advantages 2
    Supporting technologies and design considerations 3
    Mobile positioning technologies 5
    Mobile GIS architecture 6
    ArcGIS suite of web and native apps for Mobile GIS 8
    Four pillars of Mobile GIS capabilities 9
    Mobile GIS basic workflows 10
    External professional GPS receivers 11
    Introduction to Field Maps 12
    Getting started with ArcGIS Arcade 13
    Tutorial 1: Design layers, forms, and maps for a basic inventory workflow using Field Maps 14
        1.1: Create feature layers and a web map using Field Maps Designer 15
        1.2: Build the forms and design the layer schemas 16
        1.3: Improve smart forms using Arcade 20
        1.4: Style the layers and create feature templates 25
        1.5: Refine the layer and app settings 28
        1.6: Use the Field Maps mobile app to collect data 31
        Assignment 1: Design layers, forms, and a web map for data collection using Field Maps .. 35

Chapter 2: Situational awareness and one-to-many inspections using Field Maps 37
Objectives 37
Introduction 37
Situational awareness in the field 38
    Situational awareness with Field Maps 38
    Feature layer pop-ups versus forms 40
    Feature layer styles versus feature templates 40
    Geofences 41
    Data security and views 41
Tutorial 2: Design for situational awareness and inspections with related records 42
    2.1: Create a hosted feature layer from an existing dataset 43
    2.2: Configure web map layer style and pop-ups 45
    2.3: Configure layer and related table forms 49
    2.4: Configure Geofences... 53
    2.5: Configure web map bookmarks, filters, and search 56
    2.6: Collect related records using situational awareness 58
    Assignment 2: Use Field Maps for situational awareness and inspections with a one-to-many relationship 62

Chapter 3: Formcentric data collection using ArcGIS Survey123 63
Objectives 63
Introduction 63
    The need for formcentric data collection 63
    Survey123 workflow and components 64
    Survey123 Web Designer, Connect, and XLSForm basics 65
Tutorial 3: Design smart surveys and collect data using Survey123 67
    3.1: Design a basic form using Survey123 Web Designer 68
    3.2: Configure calculations and dynamic lists in Web Designer 72
    3.3: Publish your survey and review the items created 76
    3.4: Collect data using the Survey123 field app and review data using the Survey123 website 78
    3.5: Get started with Survey123 Connect (optional) 81
    3.6: Add repeats and related tables using Survey123 Connect (optional) 84
    Assignment 3: Create a smart form using Survey123 for data collection 88

Chapter 4: Rapid data collection using ArcGIS QuickCapture 89
Objectives 89
Introduction 89
    The need for data collection at speed using QuickCapture 90
    QuickCapture workflow 92
    Integration with oriented imagery 93
    Integration with drones.. 94
Tutorial 4: Design QuickCapture projects for rapid data collection and oriented
    imagery-based inspection 95
    4.1: Configure the feature layers, styles, and web map........................... 96
    4.2: Design a basic QuickCapture project 101
    4.3: Add device variables and Arcade calculations 103
    4.4: Collect data using the QuickCapture field app.............................. 106
    4.5: Inspect assets by integrating QuickCapture with oriented images............. 110
    Assignment 4: Create a QuickCapture project for rapid data collection 116

Chapter 5: Mobile workflow in offline mode................................. 117
Objectives 117
Introduction 117
    The need for offline workflows 117
    Offline map areas and mobile map packages for Field Maps...................... 118
    Copy (sideload) and reference offline basemaps 119
    Survey123 inbox and outbox... 121
Tutorial 5: Design offline workflows with Field Maps and Survey123 122
    5.1: Validate layers for offline use in Field Maps Designer........................ 123
    5.2: Create preplanned offline areas 124
    5.3: Use preplanned areas and create on-demand offline areas in the Field Maps app. 127
    5.4: Copy and use the VTPK basemap in the Field Maps app..................... 129
    5.5: Enable the inbox and link the offline basemap in Survey123 Connect (optional)..130
    5.6: Copy the VTPK basemap and use the inbox in Survey123 132
    Assignment 5: Use Field Maps for situational awareness and inspections in offline
        mode 136

Chapter 6: Workforce coordination and location sharing 137
Objectives 137
Introduction 137
       The importance of workforce coordination 138
       Workforce coordination with Field Maps Tasks 138
       Workforce coordination with the Survey123 inbox filter 139
       Location sharing for enhanced situational awareness and collaboration 140
       Key components of location sharing 141
Tutorial 6: Coordinate workforce through assignments and location sharing 143
       6.1: Configure an assignment schema and inbox query using Survey123 Connect 143
       6.2: Assign tasks and pick up tasks using Survey123 145
       6.3: Create a track view 148
       6.4: Share your locations using mobile apps 150
       6.5: Monitor, visualize, and analyze locations shared 152
Assignment 6: Create a survey or web map to support workforce coordination using assignments or tasks 155

Chapter 7: Responsive web apps for mobile devices 157
Objectives 157
Introduction 157
ArcGIS configurable responsive web app templates and app builders. 158
Basic workflow to create dashboards. 159
Basic workflow to create web experiences 160
Sources, targets, and actions 161
Considerations for mobile web apps 162
Tutorial 7: Explore and configure responsive web apps for data review 163
       7.1: Explore ArcGIS StoryMaps stories on your mobile devices 163
       7.2: Explore the desktop view of a dashboard 164
       7.3: Configure a mobile view for the dashboard 167
       7.4: Explore the desktop layout of a web experience 171
       7.5: Configure a phone layout for the web experience............................ 175
Assignment 7: Build a responsive web app for data review and editing. 179

Chapter 8: Integration with enterprise systems 181
Objectives 181
Introduction 181
    Integrating Mobile GIS with enterprise systems 181
    Webhooks 182
    Microsoft Power Automate 183
    Survey123 reports and report templates 185
Tutorial 8: Automating emails and reports and integrating with Microsoft Teams and
Microsoft OneDrive 186
    8.1: Automate Survey123 email notification using webhooks 187
    8.2: Design Survey123 feature reports 191
    8.3: Integrate Survey123 reports with OneDrive 195
    8.4: Integrate Field Maps with Teams 199
Assignment 8: Integrate Survey123 with emails and Teams using webhooks 204

Chapter 9: Virtual reality, augmented reality, and artificial intelligence 205
Objectives 205
Introduction 205
    Virtual reality 206
    Web scenes, scene layers, and 360 VR experiences in ArcGIS 207
    Augmented reality, extended reality, and mixed reality 207
    Deep learning and smart assistants in ArcGIS 209
    Generative AI and AI assistants in ArcGIS 210
Tutorial 9: Create VR experiences, use deep learning packages, and explore generative
AI in Mobile GIS 212
    9.1: Experience VR using browsers, smartphones, and headsets 212
    9.2: Collect data for your VR experience 216
    9.3: Author a web scene and a VR experience 217
    9.4: Perform photo-based inventory using an object detection model 222
    9.5: Create a photo-based inventory survey using a deep learning model 224
    9.6: Enable Survey123 smart assistant and autotranslation 227
    9.7: Explore Survey123 smart assistant and autotranslation 228
Assignment 9: Create a VR experience of a fun community 231

Chapter 10: Developing custom Mobile GIS apps 233
   Objectives 233
   Introduction 233
      Mobile app development approaches 233
      ArcGIS Maps SDKs 234
      JavaScript, HTML, CSS, and responsive web design 236
      Microsoft .NET MAUI 237
      Case studies: Delivering air quality information at your fingertips 238
   Tutorial 10: Develop a game using JavaScript and .NET MAUI 240
      10.1: Create a responsive web app using HTML and CSS 241
      10.2: Generate quiz questions using JavaScript. 244
      10.3: Monitor user tapping and evaluate user answers using JavaScript 247
      10.4: Create a responsive native app using .NET MAUI 250
      10.5: Generate quiz questions using C# 256
      10.6: Monitor user tapping and evaluate user answers using C# 259
   Assignment 10: Enhance the tutorial mobile apps' responsive layout and map interaction . . .262

Image and data credits *263*

Preface

With more than 70 percent of web traffic now originating from mobile devices, the mobile platform has become the primary entrance to the cloud and a central interface for information systems. Recognizing its pivotal role, many industries have adopted a "mobile-first" strategy in developing information systems. This is particularly true for geographic information systems (GIS), where the location-aware capabilities and widespread availability of mobile platforms are particularly invaluable. Mobile GIS has transformed how we acquire, visualize, analyze, and disseminate geospatial information, establishing itself as an indispensable component of contemporary GIS architecture. Taking advantage of the widespread availability of mobile devices, networks, and cloud computing, Mobile GIS has emerged not only as a key platform in modern GIS but as an important research frontier.

As the adoption of Mobile GIS continues to grow, so does the demand for skilled Mobile GIS professionals. This book is designed to meet this expanding need, offering a comprehensive exploration of the latest knowledge and techniques in Mobile GIS. It combines a mix of foundational principles, practical methodologies, and detailed tutorials, making it an essential resource for academic and professional environments. whether you are a GIS student, instructor, analyst, manager, consultant, or app developer, this book is tailored to enhance your skills and deepen your understanding of Mobile GIS.

Structured as 10 chapters, this book includes conceptual overviews discussing key principles and step-by-step tutorials with detailed, screenshot-supported instructions. The content spans the four pillars of mobile GIS capabilities: data capture, field awareness, integration with enterprise systems, and planning and workforce coordination. It offers a comprehensive guide to the Esri® suite of mobile technologies, including native apps such as ArcGIS® Survey123, ArcGIS Field Maps, ArcGIS QuickCapture, and browser-based apps such as ArcGIS Dashboards and ArcGIS Experience Builder. Additionally, readers can acquire advanced skills to extend Mobile GIS using ArcGIS Arcade, webhooks, HTML, CSS, ArcGIS API for JavaScript™, and ArcGIS Maps SDKs for JavaScript.

In creating this book, we have adhered to principles that emphasize both enjoyment and practicality:

- **Fun and practical:** Adopting a no-code, low-code approach, this book makes learning Mobile GIS accessible and enjoyable. With no prior programming experience required, the tutorials—drawn from real-world projects—enable readers to quickly become proficient in creating common enterprise Mobile GIS solutions.
- **Comprehensive:** This book presents Mobile GIS as an integrated system, covering everything from cloud GIS to mobile client applications, database design, user interface optimization, and online and offline workflows, including location sharing and geofencing. It provides a holistic view of creating effective Mobile GIS solutions.
- **Current and cutting-edge:** Keeping pace with the rapid evolution of Mobile GIS, this book covers newer topics, such as Arcade, webhooks, virtual reality (VR), artificial reality (AR), and artificial intelligence (AI) applications, including creating VR experiences, using deep learning models, and integrating AI assistants, preparing readers for the forefront of GIS technology.

Drawing on my extensive Mobile GIS project experience at Esri and lecturing experience at Harvard University Extension, Henan University, University of Redlands, University of Texas, Tufts University, and California State University, among other institutions, the materials and labs in this book have proven to be effective.

To learn more about this book, go to links.esri.com/GTKMobileGIS. Your feedback is welcome at links.esri.com/EsriPressCommunity. I hope this book inspires you to explore the vast possibilities of Mobile GIS and empowers you to create innovative solutions.

Acknowledgments

I extend my heartfelt thanks to everyone at Esri Press who contributed to this book. Special thanks to Catherine Ortiz for the opportunity to author this book; Claudia Naber for her exemplary project management; and Maryam Mafuri and Craig Carpenter for their thorough testing and insightful recommendations. My appreciation also goes to our editor, Carolyn Schatz, our designer, Monica McGregor, and our copyeditor, David Oberman, whose careful attention to detail and creative design have significantly elevated the quality of this book.

I am grateful to Jack Dangermond, Esri cofounder and president, for his guidance on the book's content, and to Brian Cross, Beata Van Esch, Clint Brown, and William Earnshaw for their unwavering support and encouragement. I am also thankful for the advice and enlightening discussions with Jeff Shaner, Ismael Chivite, Marika Vertzonis, Jeremy Bartley, Mark Henry, Jie Chang, Mourad Larif, Luci Coleman, Julia Levermann, Dave Crawford, Travis Butcher, Selim Dissem, and many others.

The content in this book is drawn from my extensive project experiences at Esri and my lectures at institutions such as Harvard University Extension, Henan University, University of Redlands, University of Texas, Tufts University, and California State University, among others. I am grateful to my customers and students for the opportunities to work with them and incorporate their feedback, which has significantly enhanced the content and structure of this book.

Last, and most importantly, I express my deepest gratitude to my family for their tremendous love and support.

How to use this book

About this book
Getting to Know Mobile GIS introduces the core principles and practical aspects of Mobile GIS through extensive hands-on tutorials. This book has been tested for compatibility with the ArcGIS® Online June 2024 release, with features also available in ArcGIS Enterprise 11.4.

This book is ideal for students in academic settings and professionals looking to create Mobile GIS content and projects, beyond simply using Mobile GIS apps. Prior knowledge of GIS or experience with ArcGIS Online is helpful but not required.

The chapters are designed for sequential learning, with each chapter building on the previous one. Tutorials within chapters are also progressive, so it's best to follow them in numerical order. If you encounter issues completing a tutorial, most chapters include interim data to help you proceed.

Hardware and software requirements
You will need a computer (Windows or MacOS) to complete the tutorials, although a few sections on ArcGIS Survey123 Connect require a Windows computer. Additionally, access to a web browser, an internet connection, a mobile device, and a user account that has publishing capabilities for ArcGIS Online are necessary.

Licensing the software
- **Existing credentials:** If you have credentials from your institution or organization, you may use these credentials and proceed.
- **Student-use license (available for United States users):**
  - Print textbooks (purchased in the United States) come with a code printed inside the back cover.
  - Ebooks purchased through VitalSource and labeled as *courseware* come with a license. After purchase, a code is provided. Visit links.esri.com/BookCode for help locating

this code. Some chapters may require additional products not included in this license. License activation instructions are provided at links.esri.com/GTKMobileLicense.

Tutorial data

Instructions for accessing tutorial data are provided throughout the book as needed. The tutorial data is governed by a license agreement, which can be reviewed at links.esri.com/LicenseAgreement.

Accessing additional resources

- **ArcGIS tutorial gallery:** learn.arcgis.com
- **ArcGIS Online documentation:** links.esri.com/ArcGISOnlineDocumentation
- **Esri Community:** community.esri.com
- **Esri Academy:** training.esri.com

Feedback and updates

For feedback, updates, or collaboration, visit Esri Community, where users can ask questions and share experiences. Access it directly for communications about this book, at links.esri.com/EsriPressCommunity.

Additional information is available on the book's web page at links.esri.com/GTKMobileGIS.

Chapter 1
Mobile GIS overview and mapcentric data collection using ArcGIS Field Maps

Objectives
- Grasp the Mobile GIS concept and advantages.
- Understand outdoor and indoor positioning technologies.
- Understand the four key pillars of Mobile GIS capabilities.
- Use Field Maps Designer to create feature layers and layer schemas.
- Design smart forms using Arcade.
- Create feature templates to streamline data entry.
- Collect data using the Field Maps mobile app.
- Review data using Map Viewer.

Introduction

In the post-PC era, mobile devices have become integral to both our personal lives and our professional workflows. Many industries have adopted the "mobile first" strategy, recognizing mobile platforms as the primary interface for enterprise information systems. This is especially true for geographic information systems (GIS), in which the location-aware capabilities and other advantages of mobile platforms are particularly valuable. Mobile GIS has revolutionized the way we acquire, visualize, analyze, and disseminate geospatial information, making it an essential component of contemporary GIS architecture and applications.

This chapter delves into the concepts and advantages of Mobile GIS, including mobile positioning technologies and its architecture within modern GIS. It outlines the four main pillars, or capabilities, of Mobile GIS: planning and workforce coordination, data capture, field awareness, and integration with enterprise systems. Furthermore, the chapter introduces the suite of ArcGIS® responsive web apps and native mobile apps, highlighting the Field Maps component and its basic workflow.

The tutorial section guides you through the role of a solution creator, teaching you how to create hosted feature layers, design layer schemas, build smart forms, style layers, and create feature templates using Field Maps Designer. Additionally, the tutorial positions you in the role of a mobile worker, using the Field Maps mobile app to capture data points, lines, and polygons, and reviewing the collected data through ArcGIS Online Map Viewer and a dashboard.

Mobile GIS: Concepts and advantages

The mobile platform presents a new paradigm for GIS. Mobile GIS, which refers to the use of GIS on mobile devices, originated in the mid-1990s to support field operations, such as surveying and utility maintenance. Building on the widespread availability of mobile devices, mobile networks, and cloud computing, Mobile GIS has become a crucial and widely adopted platform in modern GIS systems.

The advantages of Mobile GIS over traditional desktop GIS are as follows.
- **Mobility:** Mobile devices are wireless, allowing the extension of GIS to areas where traditional wiring is impractical or expensive. This enhances situational awareness in the field and facilitates the transfer of GIS data back to the office.
- **Location awareness:** Technologies, such as GPS, cellular networks, Wi-Fi, and Bluetooth, enable precise location tracking of mobile devices. Additional sensors such as compasses, gyroscopes, and motion sensors help determine the device's orientation, tilt, and speed.
- **Ease of data collection:** Mobile GIS eliminates the error-prone paper-based methods traditionally used in field surveys by digitizing data collection, thereby reducing costs and improving data accuracy.
- **Near-real-time information:** Mobile networks provide a live connection that enhances the temporal capabilities of GIS, enabling continuous monitoring of both spatial and temporal changes in the environment.
- **Versatile communication tools:** Mobile devices integrate with various communication methods, including voice calls, text messages, photos, videos, emails, and social networking apps, facilitating effective collaboration and communication among professionals and the public.
- **Widespread accessibility:** The ubiquity of smartphones, tablets, and smartwatches has made GIS accessible to billions, significantly expanding the user base and applications of GIS.

Although this book primarily focuses on the enterprise applications of Mobile GIS, it is important to recognize that Mobile GIS serves a broad spectrum of uses for individual consumers and enterprise organizations. Consumers frequently use Mobile GIS for everyday activities such as locating nearby points of interest, finding restaurants, navigating routes, and sharing experiences with others. Meanwhile, enterprise organizations depend on Mobile GIS for critical functions, such as data visualization, field inspections, asset management and inventory, tracking assets and field crews, conducting surveys, incident reporting, and managing parcel deliveries. This wide-ranging utility highlights Mobile GIS's important role in both enterprise and consumer applications.

Mobile GIS is a rapidly advancing field, evolving alongside edge computing, geographic information science (GIScience), and artificial intelligence (AI). It now incorporates frontier technologies, such as virtual reality (VR) and augmented reality (AR) on smartphones and wearable devices. VR offers users wearing specialized visual devices the ability to immerse themselves in 3D models, significantly enhancing their spatial understanding. AR enhances real-world views with GIS data overlays captured through a camera, enriching a user's perception of reality.

Additionally, Mobile GIS harnesses machine learning and geospatial artificial intelligence (GeoAI) to automatically detect and identify objects from photographs. Integration with large language models such as ChatGPT has led to the creation of ArcGIS Generative AI Assistants, which help compile layers, create web maps, and design smart forms specifically for mobile GIS applications. As GIS technology continues to integrate with advances in computing technologies and benefit from faster internet connections, its influence and applicability are expected to grow extensively.

Supporting technologies and design considerations

Mobile devices, mobile operating systems, and wireless communications form the foundation of Mobile GIS in the following ways.

- **Mobile devices:** This category includes smartphones, tablets, VR headsets, AR glasses, smart glasses, and smartwatches, which are essentially compact computers designed to be handheld, worn on the head, or worn on the wrist. The screen sizes of mobile devices are considerably smaller than those of desktop computers and can vary significantly.
- **Mobile operating systems:** The main operating systems include Android, iOS/iPadOS, and Linux. Specialty operating systems, such as Glass OS and Meta Horizon OS, are derivatives of Android, whereas Windows Holographic is based on Windows, and visionOS draws from iPadOS.
- **Wireless communication technologies:** Key technologies include Bluetooth, Wi-Fi, cellular networks, such as 5G, and satellite internet services, such as Starlink. Each technology varies in speed, range, and cost, serving different purposes (table 1.1). For clarity, the following discussion specifies typical formats and uses of these technologies.

Table 1.1. Comparison of speed, range, and common usage of wireless communication technologies

Wireless communication technologies	Speed	Range/coverage and common usage
Bluetooth	3 Mbps	10 meters. Facilitates communication between mobile devices and peripherals, such as GPS receivers and headsets.
Wi-Fi	Up to 9.6 Gbps (Wi-Fi 6)	100 meters, extendable. Used to build local area networks.
Cellular networks	Up to 20 Gbps (5G peak), 100 Mbps (5G average)	3-6.5 k per tower (5G). Coverage mainly in populated areas. Used for longer distance data connections, voice, and video calls.
Starlink	Between 25 and 220 Mbps, 100 Mbps average	Global coverage across most continents and oceans. Especially valuable in areas where Wi-Fi and cellular networks are unavailable.

The mobile platform offers significant opportunities for GIS, but it also presents unique challenges. The advantages of mobility come with constraints, including limited CPU speed, memory capacity, battery life, bandwidth and network connectivity, screen size, and keyboard dimensions. Although advancements in technology are mitigating some of these limitations, they still need to be carefully considered in the design and development of mobile GIS solutions.

- **Simplify user interface:** Design clear and intuitive interfaces. Avoid overcrowding the display with excessive toolbars and menus.
- **Streamline data entry:** Keep data forms simple and smart, displaying only relevant fields. Minimize text entry requirements by using drop-down lists and automated calculations wherever possible.
- **Optimize map visibility:** Use bold text labels, high contrast, and straightforward symbology to ensure maps are easily readable in outdoor settings.
- **Orient maps practically:** In navigation mode, align the map with the user's direction of movement rather than a fixed due-north orientation to enhance readability and relevance.
- **Implement offline support:** Ensure functionality in disconnected areas by preloading necessary data onto devices before entering these zones.
- **Choose appropriate development approaches:** For custom native apps, choose a specific platform or employ cross-platform programming languages to facilitate single-source code deployment across multiple platforms.

Mobile positioning technologies

The foundational element of Mobile GIS is location awareness, which has significantly evolved through various positioning technologies for outdoor and indoor environments. The importance of mobile location services was underscored by legislation such as the Enhanced 911 (E911) in the United States and similar laws globally. Before 2000, GPS technology was not commonly integrated into mobile phones, which relied primarily on cellular networks for locating callers. A tragic incident in Florida in 2001, in which a woman lost control of her car and ended up in a canal, illustrated the limitations of early mobile location technology. Unable to provide her location to the emergency operator, she tragically lost her life because the rescue units could not find her quickly. This incident highlighted the urgent need for precise mobile location services, prompting the introduction of laws mandating mobile providers to deliver the location of emergency calls with specific accuracy. these regulations have driven substantial advancements in mobile positioning technology.

Today's mobile positioning services use a combination of the following technologies (figure 1.1) to improve location accuracy, reliability, and availability.
- **Global Navigation Satellite System (GNSS):** Uses satellite signals to pinpoint precise geographic locations (longitude, latitude, altitude). Key systems include the US Global Positioning System (GPS), Russia's GLONASS, China's Beidou, and the European Union's Galileo. GNSS provides an average accuracy of 5 meters under clear skies, but this technology requires the line of sight between the GPS antenna and four or more satellites. Its accuracy can decrease because of satellite positions, cloud cover, and obstacles, such as buildings.
- **Cellular network-based positioning:** Relies on cellular network infrastructure to locate devices, using methods such as cell of origin and time difference of arrival. Although it offers broad coverage and is widely available, it is much less accurate compared with other positioning technologies.
- **Wi-Fi-based positioning:** Identifies device locations within about 100 meters of a Wi-Fi access point, with potential accuracy improvements to within 3 meters through Wi-Fi signal triangulation. This method depends on the availability of Wi-Fi signals and requires regular updates to Wi-Fi hot spot databases.
- **Bluetooth Low Energy (BLE):** Designed for low power consumption, BLE is used in tracking technologies such as Apple's AirTags, which are commonly used to attach to personal items for location sharing through apps. In buildings with dense BLE beacon setups, systems such as ArcGIS IPS™, an indoor positioning system, use algorithms to precisely locate mobile devices.

Figure 1.1. Mobile GIS uses positioning technologies to show your current location, often represented by a blue dot. This is achieved using a combination of the Global Navigation Satellite System (GNSS), cellular network–based Global System for Mobile Communications (GSM), Wi-Fi, Bluetooth Low Energy (BLE), Visual Light Communication (VLC), and other positioning technologies.

High-accuracy GPS receivers can reach submeter to centimeter location accuracy, depending on their ability to track and process satellite signals. GPS satellite signals are transmitted at different frequencies; the more frequencies a GPS receiver uses, and the more signals it receives, the more accurate it is. The accuracy can be further improved through real-time differential correction or postdifferential correction.

Mobile GIS architecture

Introducing Mobile GIS within the framework of Web GIS, which combines web technology with GIS, is crucial for appreciating its evolution and full capabilities. Initially in the mid-1990s, Mobile GIS mostly operated in a disconnected mode, in which operators loaded data onto devices each morning and, upon returning, docked the devices to transfer data back to a central computer.

However, with the rapid advancements in wireless communications, Mobile GIS is now primarily connected to the web, integrating seamlessly into Web GIS systems while still facilitating offline workflows. This connected approach allows Mobile GIS to continuously update servers with the latest field data. Conversely, web servers enhance Mobile GIS with comprehensive content and advanced analytics.

Esri® has two Web GIS products: ArcGIS Online and ArcGIS Enterprise, both offering similar functionalities (figure 1.2). ArcGIS Online is a cloud-hosted software-as-a-service

(SaaS) solution, entirely managed by Esri, which relieves organizations from the need to maintain hardware infrastructure. In contrast, ArcGIS Enterprise is a comprehensive Web GIS software suite that organizations can install and manage on various platforms, including Windows, Linux, and Kubernetes, whether on-premises or in cloud environments, such as AWS, Microsoft Azure, and Google Cloud.

Figure 1.2. Mobile GIS is becoming an important component of Web GIS architecture. Web GIS provides back-end web services, content management, access control, and a series of apps for GIS professionals who create Mobile GIS solutions and for mobile workers who primarily use mobile apps.

Today's Mobile GIS is primarily designed as a component within Web GIS, featuring a cohesive architecture that enhances its functionality and accessibility in the following ways.
- **Portal centerpiece:** The core of the architecture is a portal—a part of either ArcGIS Online or ArcGIS Enterprise—which organizes, secures, and facilitates access to geographic information products, serving as a central hub for content management.
- **Back-end capabilities:** At the back end, GIS servers are deployed, allowing users to publish geospatial services. Additionally, there is access to ready-to-use content, such as ArcGIS Living Atlas of the World, which provides a repository of geospatial web services.
- **Client-side applications**
  - **GIS professionals:** Desktop and web clients are used by GIS professionals to design data schemas, publish services, and configure apps and projects for Mobile GIS.
  - **Mobile workers:** Mobile GIS clients are used by workers in the field, who can access layers and maps live when online or use preloaded content when operating offline.

ArcGIS suite of web and native apps for Mobile GIS

ArcGIS offers a comprehensive suite of web and native mobile apps designed for both field and indoor operations, enhancing the functionality and versatility of Mobile GIS.

Responsive web apps run inside web browsers and require data connection. This category includes ArcGIS Instant Apps, ArcGIS StoryMaps℠, ArcGIS Experience Builder, ArcGIS Dashboards, and ArcGIS Hub℠. these apps are responsive to mobile screens and allow configuration for optimal mobile layouts. Although not best suited for offline use, they work well in delivering field data for situational awareness and facilitating data collection and editing with reliable internet connection. Additionally, these browser-based apps are crucial for office-based tasks, such as data review, quality control, editing, approval, and assignment processes.

Native apps are downloaded and installed directly onto your device. ArcGIS offers a series of native apps, including the following.
- **Field Maps:** Integrates several earlier products, including ArcGIS Collector, ArcGIS Explorer, ArcGIS Tracker, and ArcGIS Workforce. Field Maps serves as an all-in-one solution that supports planning, recording and sharing location, understanding, and map-centric data capture.
- **ArcGIS Survey123:** Known for its simplicity and intuitiveness, this formcentric app allows users to create, share, and analyze surveys. It features smart forms with skip logic, defaults, flexible formulas, and powerful data-pulling capabilities. The app facilitates easy data collection through the web or mobile devices in any environment, requiring minimal training.
- **ArcGIS QuickCapture:** An ideal solution for rapid data collection features a user interface with large buttons, designed to minimize interaction with the device during fast-paced field data collection. It supports offline and online work, location sharing, mapping, and data collection.
- **ArcGIS Indoors™ mobile app:** Provides an indoor mapping experience to help users navigate their organization's indoor environment. Features such as wayfinding, routing, and location-sharing enhance connectivity to the workplace or campus and integrate with calendars for seamless navigation to meetings. Users can also report incidents related to indoor assets or locations.
- **ArcGIS Mission Responder:** Part of the ArcGIS Mission product, this app allows field users to participate in missions, offering tactical situational awareness and enabling communication and collaboration through real-time messaging and reporting.
- **ArcGIS Earth:** An interactive 3D tool for planning, visualizing, and evaluating events on the globe, it offers situational awareness on desktop and mobile devices and supports data in various formats, including 3D models, KML/KMZ files, and TXT files. ArcGIS Earth allows users to quickly manipulate 3D data and collaborate for enhanced decision-making.

Four pillars of Mobile GIS capabilities

Although data capture is the most used capability, ArcGIS Mobile GIS products offer a broader range of functionalities, organized into four main pillars (figure 1.3), as follows.

- **Field awareness:** Mobile maps significantly enhance situational awareness in the field. Mobile apps and maps can be used anytime and anywhere they are needed. Field Maps includes exploratory map tools and contextual pop-ups that deliver detailed information, including charts, documents, and photos, essential for on-site decision-making. Geofencing alerts mobile workers when they enter or leave designated areas. Integrated routing, navigation, and wayfinding capabilities efficiently locate assets, whereas location sharing allows workers to view, record, and share their locations. Field Maps also supports advanced capabilities, including viewing and tracing utility networks, measuring distances, and placing points by distance along linear referenced features.
- **Data capture:** The primary use of Mobile GIS. ArcGIS provides Field Maps, Survey123, and QuickCapture, each offering unique user experiences. these apps use both indoor and outdoor locations and can integrate with external, high-accuracy GNSS receivers for sub-foot accuracy. Smart forms eliminate paper-based workflows by improving the accuracy and speed of data collection through conditional logic, pick lists, and calculated expressions. Captured photos document and automatically integrate data of various formats into forms. This functionality is based on the layers and tables in Web GIS, with device edits syncing to the enterprise GIS for immediate availability.
- **Planning and coordination:** Field Maps and Survey123 enhance field operations by supporting mobile project planning, data preparation, task or to-do list assignments, prioritizing tasks, visualizing progress, and sharing the live location of field crews. they also enable communication between the field and the office by pushing notifications and messages.
- **Integration:** The full potential of mobile GIS capabilities is realized through integration with broader organizational systems. Data captured in the field feeds into enterprise information systems, enabling analysis and dissemination of information across the organization. Tools such as ArcGIS API for Python, ArcPy™, and webhooks can automate the integration, streamline tasks, generate reports, and optimize field workflows.

Figure 1.3. Mobile GIS workflow requires four categories of capabilities, including planning and coordination, field awareness, data capture, and integration with enterprise systems.

Mobile GIS basic workflows

This book introduces a no-code/low-code approach to developing Mobile GIS solutions. The fundamental workflow spans from data layers to web maps and mobile apps.

Step 1: Design schemas and publish feature layers.

Although raster layers can be used as background on the maps, Mobile GIS primarily works with feature layers, including points, lines, and polygons. Various methods for publishing feature layers include the following:
- Use Field Maps Designer or Survey123 Designer and Survey123 Connect to create layers.
- Create empty feature layers and define attribute fields interactively in ArcGIS Online by navigating to Content > New Item > Feature Layer > Create a Blank Layer and selecting the required geometry type. After creating the layer, proceed to the item page > Data > Fields to add fields.
- Create feature layers from existing templates or by copying the schemas of existing layers.

- Upload your own data, such as CSV files, zipped shapefiles, GeoJSON, or zipped file geodatabases. Navigate to ArcGIS Online Content > New Item, select From Your Computer if your data is stored locally, or choose From a Cloud Drive if it's stored on services, such as Google Drive, Dropbox, or OneDrive.
- Publish feature layers using ArcGIS Pro, a desktop GIS application that enables GIS professionals to develop complex data models, such as utility networks, and publish feature layers. these layers can be hosted, with data managed by ArcGIS, or referenced, with data remaining in your enterprise geodatabase and managed by your organization.

Step 2: Configure layers and web maps.
- Configure the layer to be read-only or enabled for editing and configure the type of editing—that is, what the editor can see and edit.
- Create view layers to allow user groups to have appropriate permissions.
- Configure layer styles and pop-ups for viewing the layers.
- Configure feature templates and forms for data collection.
- Refine web map settings such as the offline capabilities, Geofence settings, search, and filter.

Step 3: Perform field capture and situational awareness tasks.
- Access the web map, layers, and forms as a mobile worker.
- Copy (sideload) data to mobile devices if needed.
- View maps, query data, and collect data.

External professional GPS receivers

For Mobile GIS applications that require high horizontal and vertical accuracy, such as land surveying and utility mapping, external professional-grade GNSS receivers or GPS receivers are crucial. these receivers achieve submeter to centimeter-level accuracy, significantly surpassing the 3- to 5-meter accuracy of standard built-in mobile device GPS.

GNSS refers broadly to all satellite navigation systems. Although GPS is the most widely recognized, GNSS also includes systems such as GLONASS (by Russia), BeiDou (by China), and Galileo (by the European Union). The term "GPS" is often used interchangeably with "GNSS" because of its widespread recognition. Professional GNSS receivers can access more satellites and leverage multiple systems, enabling more precise location calculations and altitude determinations by accessing a greater number of satellites simultaneously. The accuracy of GNSS receivers improves with the number of frequencies and satellite signals processed.

These receivers often support real-time differential correction technologies, such as satellite-based augmentation systems (SBAS) or real-time kinematic (RTK), which enhance positional accuracy instantaneously and ensure the reliability of collected spatial data.

In environments where signal disruption is common—such as dense urban settings, forests, or mountainous areas—professional-grade GNSS receivers can maintain strong signal reception. This ensures continuous and reliable data collection even where ordinary mobile receivers might fail.

ArcGIS Field Maps and Survey123 can use professional-grade GPS receivers that support standard National Marine Electronics Association (NMEA) sentence outputs. For iOS devices, the receivers must be specifically supported by iOS. Connections to these applications are typically made through Bluetooth or USB, allowing users to enhance data collection and significantly improve the quality of spatial data. This capability is vital for mobile workers in industries that depend on accurate, real-time geospatial data. Moreover, Field Maps and Survey123 can capture detailed GPS metadata, such as signal strength, the number of satellites used, and positional precision (HDOP/VDOP), which is crucial for evaluating data quality and ensuring compliance with data collection standards.

Introduction to Field Maps

Field Maps is an all-in-one app that is widely used for mapcentric data collection, functioning with or without network connectivity. It also supports utility network editing and tracing, location sharing, Geofences, task assignments, workforce coordination, webhook automation, and more. Field Maps has two main components (figure 1.4).

- **Field Maps Designer:** This component is for solution creators or map authors to use. It can design layers and maps, create smart forms, enable maps for offline use, set up Geofences, and configure various settings to optimize maps for field workflows.
- **Field Maps mobile app:** This component is for mobile workers to use. Using this app, mobile workers can open the maps, forms, data, and tools necessary to complete their field tasks. Field Maps is designed to function online and offline, as well as in outdoor and indoor environments.

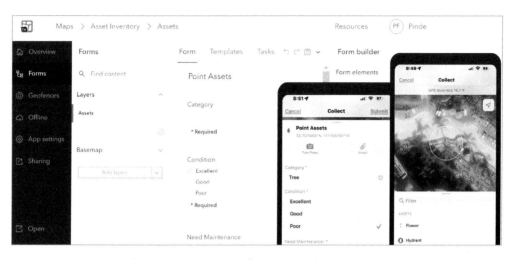

Figure 1.4. ArcGIS Field Maps Designer (*left*) and mobile app (*right*).

Getting started with ArcGIS Arcade

Arcade is a portable, lightweight, and secure expression language tailored for creating custom content in ArcGIS applications. Its syntax closely resembles JavaScript, allowing users to perform mathematical calculations, format text, and evaluate logical statements. Arcade stands out among scripting languages for its unique ability to process feature and geometry data types. Arcade is supported across the ArcGIS system. For example, the Arcade scripts you author for pop-ups in one app, such as Map Viewer, web apps, mobile apps, or ArcGIS Pro, are supported in the rest of the apps.

Arcade offers various profiles suited for different purposes. This book will cover the form calculation, pop-up, and QuickCapture profiles. This chapter begins with the Form Calculation profile, which is used for authoring calculated expressions similar to Microsoft Excel formulas but more sophisticated because of Arcade's capabilities. these expressions can automatically calculate and update attribute values in Field Maps forms and are respected across various platforms, including the Edit tool of Map Viewer, the Edit widget of Experience Builder, and ArcGIS Web Editor.

The form calculation profile comes with a built-in variable $feature, representing the current feature being added or edited, providing access to its attributes and geometry of the features. For example:
- To retrieve the value of an attribute field named AGE_5DOWN:
    $feature.AGE_5DOWN or $feature["AGE_5DOWN"]
- To calculate and round the percentage of the young in the total population:
    Round($feature.AGE_5DOWN / $feature.TOTAL_POP)

- To access the feature set from a feature layer named USA Cities in the current web map, including the geometries and the values of the attribute field NAME:
  ```
  var cities = FeatureSetByName($map, "USA Cities", ["NAME"]);
  ```
- To find the cities that intersect the current features being added or edited:
  ```
  var intersectedCities = Intersects($feature, cities)
  ```

Tutorial 1: Design layers, forms, and maps for a basic inventory workflow using Field Maps

In this tutorial, you will design a Mobile GIS solution to inventory assets in a specified area, such as your campus, community, or a local park. This tutorial will guide you through creating a functional and secure Mobile GIS solution that enhances data integrity and user engagement, ensuring that each layer and form is specifically designed to streamline the data collection process while securing the data against unauthorized modifications.

Requirements:
- **Data collection capabilities:**
  - Ability to collect point, line, and polygon assets.
  - Allow photo attachments for each type of asset.
- **Layer configuration:**
  - Each layer should have at least a category attribute field with a list of coded values.
  - Each layer must offer more than one feature template.
- **Form design:**
  - At least one layer's form must include two or more of the following: combo boxes, radio buttons, switches, or their combinations.
  - The form should be smart, incorporating one conditional visibility, one conditional requirement, and one calculated expression.
- **Access and security:**
  - Allow mobile workers to access and collect data using your web map and layers in Field Maps.
  - Configure the layers to prevent any mobile worker from modifying data collected by others.
- **Data collection:**
  - Collect some features using Field Maps mobile app to test your forms and web map.

1.1: Create feature layers and a web map using Field Maps Designer

1. Start a web browser, navigate to ArcGIS Online (https://www.arcgis.com, or the URL to your ArcGIS Online organization), and sign in.

   If you don't know the URL to your ArcGIS Online organization, check with your GIS administrator or instructor.

2. In the upper right of the page, click the app launcher and select Field Maps Designer.

   The Maps page displays all the maps you own or have privileges to edit.

3. Click the New Map button.

4. On the Create Layers tab, apply the following settings:
   a. For Layer name, replace Layer_1 with Point Assets. Keep the Layer type as Point Layer.
   b. Click the Add button to add another layer.
   c. For Layer Name, replace Layer_1 with Line Assets. For Layer Type, choose Line Layer.
   d. Click the Add button to add another layer.
   e. For Layer name, replace Layer_2 with Polygon Assets. For Layer Type, choose Polygon layer.
   f. Click Next.

   The Layer Settings tab appears. these settings allow you to configure tasks, high-accuracy GPS metadata fields, and z-value and m-value enabled geometries. Tasks are discussed in chapter 6. You don't need to change these default settings.

   Next, you will define the layer's coordinate system.

5. On the Layer Settings tab, apply the following settings:
   a. Click Advanced settings, expand the Set Coordinate System drop-down list, choose Use a Coordinate System, and click Browse.
   b. In the Coordinate Systems window, search for Web Mercator.
   c. In the search result, choose WGS 1984 Web Mercator (auxiliary sphere).
   d. Click Done.
   e. Click Next.

The Title and Save tab appears. Here you will set the title of the web map, which is the map name that your mobile workers will see, and the title of the host feature layer, which will include the point, line, and polygon assets sub-layers you defined above.

6. On the Title and Save tab, apply the following settings:
    a. For Map Title, type Asset Inventory.
    b. For Feature Layer Title, type Assets.
    c. For Folder, click Create New Folder, type Chapter1, and press Enter.
    d. Click Create Map.

The feature layer title needs to be unique in your ArcGIS Online organization. If the name is not available, you may append your name or initials to make the title unique.

It will take a moment for the map and layers to be created. After they are created, the form builder becomes available. You will build the forms in the next section.

1.2: Build the forms and design the layer schemas

In this section, you will design a form for each of the layers, and this will create the corresponding attribute fields for the layers.

> **Note:** When editing the map, layers, and form in Field Maps Designer, avoid editing them simultaneously in separate browser tabs. Editing these elements across multiple tabs can result in data loss or data being overwritten.

1. In the Forms pane, under Layers, make sure the Point Assets layer is selected.

    If you have closed the browser, go to ArcGIS Online, sign in, click the app launcher, click Field Maps Designer, and click the map you just created.

    The layer has a blank form canvas. You will add the category, condition, need maintenance, notes, city, and inspector name elements to the form and create the fields in the layer.

2. In the Form builder pane, drag a Combo Box element to the form.

3. In a different web browser tab, open https://arcg.is/11i9qb0 to download the provided CSV file and close the Browser tab.

> **Note:** Short URLs are case-sensitive.

You will be using the download PointAssetCategories.csv file in the next step.

4. In the Form builder pane, in the Properties pane, apply the following settings:
   a. In the Formatting section, for Display Name, type Category. The Field name will be autofilled as the Display name.

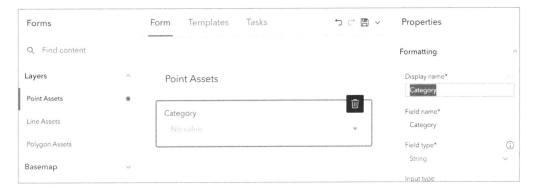

   b. Click Create List.
   c. Click Select from Your Device and browse to and select the PointAssetCategories.csv file you just downloaded.
   d. Review the labels and codes loaded.
   e. Click Done.
   f. If prompted with a dialog box asking if you want to continue, click Continue.
   g. Uncheck the Include "No value" option.
   h. Under the Logic section, check Required.
   i. Close the Properties pane.

5. Click the Save to Map button.

In the rest of the section, click Save often to avoid loss of your settings.

6. In the Form builder pane, drag a Radio Buttons element to the form, under the Category element.

7. In the Properties pane, apply the following settings:
    a. In the Formatting section, for Display Name, type Condition.
    b. Click Create List.
    c. Type three labels: Excellent, Fair, and Poor, with each code the same as the label.
    d. Click Done and click Continue if prompted.
    e. Uncheck the Include "No value" option.
    f. In the Logic section, check Required.
    g. Close the Properties pane.

Next, you will add the Need Maintenance element.

8. In the Form builder pane, drag a Switch element to the form, under the Condition element.

9. In the Properties pane, apply the following settings:
    a. For Display Name, type Need Maintenance.
    b. Under Switch Values, for Off Value, type No.
    c. For On Value, type Yes.
    d. For Default Value, choose No.
    e. Close the Properties pane.

Next, you will add a multiline text and two single-line text elements.

10. In the Form builder pane, drag a Text – Multiline element to the form, under the Need Maintenance element.

11. In the Properties pane, apply the following settings:
    a. For Display Name, type Notes.
    b. Close the Properties pane.

12. In the Form builder pane, drag a Text – Single Line element to the form, under the Notes element.

13. In the Properties pane, apply the following settings:
    a. For Display Name, type City.
    b. Close the Properties pane.

14. In the Form builder pane, drag a Text – Single Line element to the form, under the City element.

15. In the Properties pane, apply the following settings:
    a. For Display Name, type Inspector.
    b. Close the Properties pane.

16. Click Save to Map.

Next, you will configure the line and polygon assets layers by adding a Category element.

17. In the Forms pane, under Layers, click Line Assets.

18. In the Form builder pane, drag a Combo Box element to the form.

19. In the Properties pane, apply the following settings:
    a. For Display Name, type Category.
    b. Click Create List.
    c. Type three labels: Fence, Trail, and Electric Line, with each code the same as the label.
    d. Click Done and click Continue if prompted.
    e. Close the Properties pane.

20. Click Save to Map.

21. In the Forms pane, under Layers, click Polygon Assets.

22. Similarly, add a Combo Box element, with Category for Display Name, and Pond, Lawn, Sports Field for Labels and Codes.

23. Click Save to Map.

You have built a basic form for each asset layer. Through the process, you have also defined the schemas for the layers. You will review the schemas later in this tutorial.

1.3: Improve smart forms using Arcade

In this section, you will improve the forms you created in the previous section by adding conditional visibilities, conditional requirements, and calculations.

1. Under Layers, click Point Assets.

   The form for the Point Assets layer displays. You will configure the Need Maintenance field as visible when the condition is fair or poor, and as required only when the condition is poor.

2. Click the Need Maintenance element.

3. In the Properties pane, under the Logic section, next to the Required check box, click the Settings button.
     a. Click New Expression.
     b. In the Expression builder window, set the Title as Poor Condition.
     c. Choose the Condition field, the is operator, and the value Poor.
     d. Observe that the following Arcade syntax is built for you automatically:
        `DomainName($feature, "Condition") == "Poor"`
     e. Click Done.

   The Arcade line of code checks if the Condition attribute value of the current feature being collected is labeled Poor. Because the labels and the codes of the Condition attribute are the same, the code can evaluate the code—in other words, the attribute value—directly, and thus the line of script can also be written as `$feature["Condition"] == "Poor"`, or yet another way as `$feature.Condition == "Poor"`.

4. In the Properties pane, under the Logic section, next to the Visible check box, click the Settings button.
   a. Click New Expression.
   b. In the Expression builder window, set the Title as Fair or Poor Condition.
   c. Choose the Condition field, the is operator, and the value Fair.
   d. Click Add Condition.
   e. Choose the Condition field, the is operator, and the value Poor.
   f. Observe the Arcade syntax built looks like the following:
   ```
   DomainName($feature, "Condition") == "Fair" ||
   DomainName($feature, "Condition") == "Poor"
   ```
   g. Click Done.

Because the labels and the codes of the Condition attribute are the same, the script can also be written as follows: `$feature["Condition"] == "Fair" || $feature["Condition"] == "Poor"`, or `$feature.Condition == "Fair" || $feature.Condition == "Poor"`.

Next, you will configure the Notes element to be visible only when Need Maintenance is set to Yes.

5. Click the Notes element.

6. In the Properties pane, next to the Visible check box, click the Settings button.
   a. Click New Expression.
   b. In the Expression builder window, set the Title as Maintenance Needed.
   c. Choose the Need Maintenance field, the is operator, and the value Yes.
   d. Observe the Arcade syntax built looks like the following:
   ```
   DomainName($feature, "Need_Maintenance") == "Yes".
   ```
   e. Click Done.

Next, you will calculate the inspector's full name, so that the mobile workers won't have to type their names manually.

7. Click the Inspector element.

8. In the Properties pane, under the Logic section, next to Calculated Expression, click the Settings button.
   a. Click New Expression.
   b. In the Expression builder window, set the Title to Calc Full Name.
   c. Click the Functions button (fx) on the right side of the window. Search for GetUser. In the search result, notice the multiple ways to get user info.

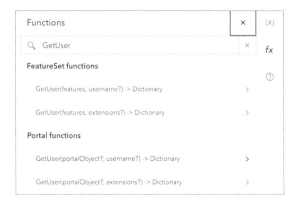

   d. Under Portal Functions, notice GetUser(portalObjects?, username?). Click the arrow button on the right to learn more about it.
   e. Read the description of the function and understand the parameters and the return value. Scroll down to review the examples.
   f. In the script area, type the following code or copy it from this page: https://arcg.is/0enOK4.
      `GetUser(Portal('https://arcgis.com')).fullName`
   g. Click Run and notice your full name displays in the Output area.
   h. Click Done.

Next, you will calculate the city that the asset location is within. You will add the cities layer to the map so that Arcade can access the layer more easily.

9. Click Save to Map.

10. On the Contents toolbar, click Open and choose Map Viewer.

Map Viewer opens with your layer listed. Next, you will add a provided cities layer.

Map Viewer toolbars

Map Viewer has two vertical toolbars on the left and right sides of the view. The dark toolbar on the left is the Contents toolbar. This is used for managing layers and tables. The light toolbar on the right is the Settings toolbar. This is used to access options for configuring and interacting with map layers. there are Expand and Collapse buttons at the bottom of each toolbar to expand or collapse them.

11. In Map Viewer, apply the following settings:
    a. On the Contents toolbar, click the Add button.
    b. Above the search bar, click My Content to expand the list. Choose ArcGIS Online.
    c. Search for USA Cities owner:GTKMobileGIS.

    d. Click Add to add the found layer to the map.
    e. Click the back arrow to go back to the Layers list.
    f. Drag the USA Cities layer down to the bottom so that the polygon layer won't block other layers.
    g. Zoom in to see the Cities layer display on the map.
    h. Click a city to see the attributes in the pop-up. Notice it has a Name field.

Next, you will change the basemap to Imagery Hybrid, which can serve as a better reference for placing asset locations.

12. On the left toolbar, click the Basemap button and select Imagery Hybrid.

13. Click Save and Open. Click Save.

14. Close Map Viewer.

You are directed back to Field Maps Designer. Your web map is reloaded automatically to ensure that you work with the latest of your web maps.

15. On the form for the Point Assets layer, click the City element.

16. In the Properties pane, click the Settings button next to Calculated Expression.
    a. Click New Expression.
    b. In the New Arcade Expression window, for Label, type Calc City.
    c. In the script text area, type the following code or copy it from this page: https://arcg.is/0enOK4.
    ```
    var cities = FeatureSetByName($map, "USA Cities", ["NAME"]);
    var city = First(Intersects($feature, cities));
    if (!IsEmpty(city)) {
      return city["NAME"];
    } else {
      return null;
    }
    ```
    d. Click Run and notice the output is null.
        This is because the layer doesn't yet have features.
    e. Click Done.

The script first gets the USA Cities layer, along with the geometries and the Name field from your web map, and then uses the current asset point to intersect with the cities polygon layer. If there is an intersecting city, the script returns the city's Name attribute value; otherwise, it returns null.

17. Click Save to Map.

You have authored several of the most used Arcade scripts! these scripts have made your form simpler and more automatic.

1.4: Style the layers and create feature templates

> **Note:** At the time this book was written, the process for creating feature templates was expected to be streamlined in upcoming ArcGIS Online releases. Check https://arcg.is/1K0f5S0 for updated instructions.

In this section, you will first style the three asset layers in Map Viewer and then create feature templates in Field Map Designer.

1. In the lower left of the page, click Open and choose Map Viewer.

   Map Viewer opens with your layer listed.

2. With the Point Assets layer selected, click the Styles button on the right toolbar.

3. In the Styles pane, apply the following settings:
   a. For Choose Attributes, click Field, choose Category, and click Add.
   b. For Pick a Style, click the Style Options button of Types (unique symbols).
      Each of the categories gets a unique color.
   c. Click Done and click Done again.

   To keep the tutorial flowing smoothly, you can use these simple styles. Feel free to explore the available symbols and enhance the styles—for example, by using icons instead of just colors.

4. On the right toolbar, click the Properties button.

5. Click Information to expand it and click Save.

   Clicking the Save button here will save the layer's style to the layer itself, more specifically in the layer definition.

   Next, you still style the Line Assets layer similarly.

6. In the Layers pane on the left side of the map, click the Line Assets layer.

7. On the Settings toolbar, click Styles.

8. In the Styles pane, apply the following settings:
    a. For Choose Attributes, click Field, choose Category, and click Add.
    b. For Pick a Style, click the Style Options button of Types (unique symbols).
    c. Click Done and click Done again.

9. On the Settings toolbar, click the Properties button.

10. Click Information to expand it and click Save.

Next, you still style the Polygon Assets layer similarly but will change the default transparency.

11. In the Layers pane on the left side of the map, click the Polygon Assets layer.

12. On the Settings toolbar, click Styles.

13. In the Styles pane, apply the following settings:
    a. For Choose Attributes, click Field, choose Category, and click Add.
    b. For Pick a Style, click the Style Options button of Types (unique symbols).
    c. For Symbol style, click the Edit button (pencil).

    d. In the Symbol Style pane, set the Fill transparency to 80%.
    e. Click Done and click Done again.

14. On the Settings toolbar, click Properties.

15. Click Information to expand it and click Save.

16. On the Contents toolbar of Map Viewer, click Save and Open and then click Save.

17. Close Map Viewer.

You are directed back to the Forms page of Field Maps Designer. Next, you will create the feature templates for each layer.

18. In the Forms pane, click the Point Assets layer to select it.

19. Click the Templates tab and notice that there is only the New Feature template.

    Next, you will create feature templates based on the category field, which contains 10 unique values. Duplicate the current New Feature template nine times for a total of 10 templates.

20. Hover over the New Feature template and click the Duplicate button nine times.

    Next, you will click each of the templates to set its category value and display name.

21. While the New Feature template is selected, in the Properties pane, under Default Values, click Category and choose Bench. Under formatting, set the Display name to Bench.

22. Repeat the process for the rest of the templates. Set their default Category values and Display names to each of the asset categories. As you do so, notice the symbol of each category display.

23. Click Save.

    Similarly, you will create the feature templates for the Line Assets layer. The Category field has three unique values.

24. On the Forms pane, click the Line Assets layer to select it.

25. Click the Templates tab.

26. Hover over the New Feature template and click the Duplicate button twice.

27. Click each of the templates and set their default Category values and Display names to each of the asset categories.

28. Click Save.

29. Repeat the process and create three feature templates for the Polygon Assets layer.

30. Click Save.

    The feature templates configured in this section will make it easier for your mobile workers to select the correct asset category and automatically set the corresponding category value for each one.

1.5: Refine the layer and app settings

In this section, you will refine the layer settings for better security and the app settings for ease of use.

1. Start a web browser, navigate to ArcGIS Online (www.arcgis.com, or the URL to your ArcGIS Online organization), and sign in.

2. On the top menu, click Content.

3. Under Folders, click All My Content.

    Your content items are listed.

4. Click the Assets feature layer.

    The layer's item page displays and lists the point, line, and polygon assets sublayers.

5. Under Layers, click the Point Assets layer.

6. On the right side of the page, under Attachments, notice that the layer has Enable Attachments selected.

    Using this setting, mobile workers can upload photos, videos, PDFs, and other documents when capturing data. The line and polygon assets layers also have attachments enabled.

7. In the upper left of the page, click the back arrow to go back to the parent layer.

8. Click the Settings tab.

9. In the General section, under Public Data Collection, choose Approve This Layer to Be Shared with the Public When Editing Is Enabled, and click Save.

This setting makes it easier for your instructor or colleagues to access and use your layer. In real projects, you typically wouldn't make your layer editable by all users.

10. In the Feature Layer (Hosted) section, under Editing, review the following settings:
    a. Notice that this layer is editing enabled, which is why your mobile workers can use this layer to capture data.
    b. Notice the option Keep Track of Changes. Leave as is.
       When enabled, this setting records changes to the feature layer in a system change log. By default, changes are retained for 180 days, allowing you to restore data during this period, such as identifying and reinstating deleted records.
    c. Notice the option Keep Track of Who Edited the Data is selected. This adds four editor tracking fields to your layer: CreationDate, Creator, EditDate, and Editor. The field names may vary.
    d. Notice the Enable Sync option. This option is needed for offline support, and it adds the GlobalID field to your layer.

11. For the question What Features Can Editors See, choose Editors Can Only See their Own Features.

    This setting enhances the security of your layer by preventing mobile workers from deleting or editing other workers' data. In subsequent chapters, you will use view layers to further enhance the security of your data.

12. Click Save.

13. Click the Data tab on the top and review the table view of your data.

    The Point Assets layer's attribute table displays. You layer is empty for now.

14. Click the Fields tab and review the following fields of the Point Assets layer:
    a. CreationDate, Creator, EditDate, and Editor are editor tracking fields.
    b. GlobalID field.
    c. The fields that were created by Field Maps Designer.

15. Click the Category field and review the List of Values (Domain).

    As you were designing the form for the layer, the elements were saved as fields in the attribute table.

16. In the upper left, note the layer drop-down list. Click the list to select the Line Assets or the Polygon Assets layer to review their schemas.

    This book will use metric measurements throughout, so it is recommended that you change your measurement settings.

17. In the upper right, click your profile picture and click View Profile. On the left, click View My Settings. Scroll down to the Units section and choose Metric.

    Next, you will review the app settings of the map in Field Maps Designer.

18. Click the app launcher and choose Field Maps Designer.

19. Click the Asset Inventory map.

20. On the Contents toolbar, click App Settings.

21. Click Collection to expand it.

22. Increase the default location accuracy requirement from 10 meters to 30 meters.

23. In the upper-right corner of the page, click Save Changes.

    For indoor environments such as classrooms or offices without an indoor positioning system, the location accuracy of your phone or tablet may be low. Setting the accuracy threshold to 30 meters or more will allow you to experiment with data capture indoors in the next section.

    For real-world projects, you may require higher location accuracy by reducing the threshold to a smaller value to ensure the quality of your data.

24. On the Contents toolbar, click the Sharing button.

25. Click Set Sharing Level, choose Everyone, and click Save.

26. Under Sharing Options, notice the URL and QR code options.

    In the next section, you will use one of the two options to open the map on your mobile device.

1.6: Use the Field Maps mobile app to collect data

In this section, you will use Field Maps on your mobile device to collect data. The tutorial is based on the iPhone edition of Field Maps, but other editions will be similar. If possible, consider taking a walk around your campus, community, or a local park to capture some assets.

1. Install ArcGIS Field Maps on your mobile device if you haven't done so.

   Field Maps is available from Google Play for Android devices and the App Store for Apple devices.

2. Tap the Field Maps app to open it. Sign in with your ArcGIS Online account.

3. Find and open the Asset Inventory web map you created in the previous section. Optionally, you can use a similar web map with improved layer styles by scanning the QR code below on your mobile device.

4. If you are prompted to allow ArcGIS Field Maps to access your location, click Allow While Using App.

   You can also enable location services for Field Maps manually. For example, for iOS, go to Settings > Privacy & Security > Location Services > Field Maps > While Using the App and enable Precision Location.

   The map appears and automatically centers on your current location by default, which is convenient if you are at the location of the asset.

5. At the top of the map, review the GPS accuracy information.

If you are in an open space, the accuracy can be within a few meters. However, if you are indoors without IPS and away from windows, the accuracy may be considerably reduced. This is why you relaxed the accuracy requirements in the previous section.

6. Tap the Add button to collect a new asset.

The feature templates display the available asset categories along with each category's symbol.

7. Use your current location and tap an asset category from the feature template.

8. Manually improve the location by zooming and panning the map. Tap the Update Point button.

9. Try the Fair and the Poor conditions and observe the smart form behavior:
    a. When Excellent, there are no Need Maintenance or Notes questions.
    b. When Fair, the Need Maintenance question becomes visible but is not required.
    c. When Poor, the Need Maintenance question is required.
    d. When Need Maintenance is selected, the Notes question becomes visible.

The Inspector and City questions are calculated automatically by the Arcade scripts you authored in the previous section.

10. Tap the Take Photo button.

11. If prompted that Field Maps wants to access the camera, click OK.

12. Take a photo and tap Use Photo.

The photo appears in the form. You can take multiple photos. You can also tap the Attach button to select and attach videos, audios, or other types of files.

13. Tap Submit to save the asset.

Next, you will collect another point asset.

14. Press and hold the map at the location of an asset you want to capture.

A pin is dropped on the map.

Chapter 1: Mobile GIS overview and mapcentric data collection using ArcGIS Field Maps   33

15. On the Information pop-up, choose Collect Here.

    The location target appears on the map at the pin's location.

16. Fill in the attributes, optionally attach a photo, and submit the asset.

    Next, you will collect a line asset.

17. Tap the Add button again. Choose a Line Asset type.

18. Choose one of the following two ways to add the line:
    a. If you are indoors, you can manually specify the line. Zoom and pan the map to your starting point. Tap the Add Point button. Move the map and add more points to complete the line.
    b. If you are outside, you can collect the line using streaming, which captures the line as you walk along it. Tap the Options button (three dots) and tap Start Streaming. Walk along the line you want to collect. The line is drawn on the map as you walk. Tap Stop Streaming to finish the line.

    The streaming interval setting controls how often or how far apart the vertices are added to the line. This can be configured in Field Maps Designer app settings or in the mobile app.

19. Tap Submit to save the line asset.

    You just collected several assets using Field Maps. The collected data is saved to the feature layers you created in the previous section.

20. Similarly, collect a polygon asset.

21. Capture more assets as you want.

    You have collected several assets. Assuming you have an internet connection, the data has already been uploaded to the cloud and saved to your feature layer.

    Next, you will review the data you have collected.

22. Open a web browser on your computer.
    a. If you used your own web map, navigate to ArcGIS Online, sign in, go to your content, find the web map you created in section 1.1, and open it in Map Viewer. Click an asset on the map.
    b. If you used the web map provided, navigate to https://arcg.is/0qOjTb and click an asset on the map, or navigate to https://arcg.is/1b1S1G0 and click an asset in the table.

23. Observe the location of the asset on the map. Observe its attributes and attachments in the pop-up.

24. Repeat the process to review more assets.

In this tutorial, you assumed two roles:
- First, as a map author creating mobile solutions using Field Maps Designer. You developed a feature layer comprising three sublayers: point, line, and polygon. Your design integrated smart forms and schemas for each layer, bolstered by conditional visibility, requirements, and calculated expressions using Arcade. This configuration simplified user interactions, minimized manual data entry, and enhanced data accuracy. You also designed custom layer symbols and feature templates to streamline operations and maintain data quality.
- Second, as a mobile worker collecting data in the field with the Field Maps mobile app. You captured data for points, lines, and polygons, either individually or through GPS streaming. Using feature templates allowed you to automatically set specific attribute values. You effectively navigated the smart forms, confirming the logic's effectiveness; you chose most field values from predefined lists to reduce typing, saw only relevant fields, addressed mandatory questions, and verified the value calculations performed by Arcade. You also attached photos and other files to enrich the dataset.

The tutorial wrapped up with a review of the collected data using Map Viewer and a dashboard, showcasing the practical capabilities of Field Maps. This foundational experience prepares you to delve into more advanced features of Field Maps for enhanced data capture and field awareness in the next chapter.

Assignment 1: Design layers, forms, and a web map for data collection using Field Maps

Apply the concepts and skills covered in this chapter to develop a robust and secure mobile data collection solution for utilities inspection, park evaluation, disaster damage assessment, or other applications. Make sure each layer and form are thoughtfully designed to ease the data collection process while ensuring the integrity and security of the data collected.

Requirements:
- **Create a Point feature layer:**
  - Must include at least three fields.
  - Attachments must be enabled.
  - The form should feature at least two of the following: combo boxes, radio buttons, switches, or their combinations.
  - Include at least one conditional visibility, one conditional requirement, and one calculated expression in the form.
  - The layer must have more than one feature template.
- **Create a Line or Polygon feature layer:**
  - Include at least one attribute field.
  - The layer should have more than one feature template.
- **Access configuration:**
  - Configure the layers such that one mobile worker cannot modify data collected by others.
  - Allow your instructor to open your web map and layers in Field Maps and collect data.
- **Data collection:**
  - Collect five or more features for each of your two layers.

What to submit:
- The URL or QR code to your web map.

Chapter 2

Situational awareness and one-to-many inspections using Field Maps

Objectives

- Understand the importance of situational awareness in field settings using Mobile GIS.
- Differentiate between layer pop-ups and forms, as well as styles and feature templates.
- Publish feature layers from existing datasets.
- Create feature layer views to improve data security.
- Configure styles and pop-ups for situational awareness in the field.
- Implement smart forms using Arcade.
- Create and configure Geofences.
- Explore the situational awareness capabilities of Field Maps.
- Work with one-to-many relationships in Field Maps.

Introduction

Building on the previous chapter in which you created new feature layers for data collection from the field to the office, this chapter shifts focus to publishing your existing data to facilitate situational awareness in the field. This process includes understanding the locations and attributes of existing features; receiving alerts when entering and exiting designated areas; and employing various functions, such as querying, filtering, navigation, and analysis, to locate and navigate to features.

This chapter delves deeper into situational awareness and enhances data collection processes for a typical inspection workflow involving related tables. It begins by publishing an existing dataset of hydrants and related inspections to ArcGIS Online as both a feature layer and a table. Key tasks include designing layer styles and pop-ups, as well as configuring web map searches and filters to improve situational awareness. Additionally, this chapter covers the configuration of forms on layers and related tables using Arcade scripting to optimize data collection and integration. The section then transitions to the mobile app, demonstrating how to use the previously configured capabilities for enhanced situational awareness and efficient data collection. This comprehensive approach not only streamlines the workflow but also ensures that field personnel are equipped with the necessary tools and information for effective field operations.

Situational awareness in the field

Field operations personnel rely heavily on Mobile GIS to maintain situational awareness. Essential to their tasks is the ability to access detailed information about assets or layers they need to inspect, operate, or maintain. This includes knowing the current locations, conditions, other attributes, and attached photos and other documents. Additionally, having a record of past inspections or maintenance performed on these assets is often important.

Mobile workers also need access to various reference layers, such as demographics, soil types, terrain features, and imagery, to gain a comprehensive understanding of the environments they are working in. Awareness of any potential hazards or safety concerns in their working areas is critical; for example, they must be cautious of any environmental constraints or regulatory compliance requirements in specific zones. Performing analysis is another key function of Mobile GIS in field operations. For instance, workers might need to trace a utility network to determine which valves should be shut off to stop water from bursting or to prevent gas leaks escaping from a broken pipe.

The availability of this information and the necessary tools, both online and offline, is paramount for mobile workers to complete their tasks effectively while avoiding and mitigating safety hazards. The capabilities of Mobile GIS ensure that field operations are conducted with the highest levels of efficiency and safety.

This chapter, along with chapters 5, 6, 7, and 9, discusses how ArcGIS products support situational awareness through various applications. these include responsive web apps, such as Experience Builder and Dashboards (detailed in chapter 7); offline capabilities (chapter 5); location sharing (chapter 6); and virtual and augmented reality technologies (chapter 9). The focus of this chapter is on using Field Maps in online mode.

Situational awareness with Field Maps

Field Maps serves as a comprehensive tool for enhancing situational awareness and operational efficiency in the field. It facilitates the creation and sharing of web maps specifically configured for field crews and their tasks. these maps consolidate all necessary data into clear, accessible formats available online and offline, enabling teams to access critical information on any device, anywhere (figure 2.1). Key capabilities include the following.

- **Diverse data layers and rich information:** Field Maps supports a variety of map layers, layer styles, pop-ups, charts, and rich text, making it highly configurable. Although this chapter emphasizes feature layers, Field Maps also accommodates map image layers, raster tile and vector tile layers, utility network layers, and mobile map packages.

- **Geofencing:** This capability enables the configuration of zones of certain polygons and buffers. Field crews receive push notifications upon entering or exiting these areas, enhancing safety and compliance by providing details about potential hazards and regulations.
- **Search and filter functions:** Both search and filter allow users to query for specific features by attributes. Search highlights the found features whereas filter displays only the matching features.
- **Navigation tools:** Field Maps provides essential navigation assistance, helping mobile workers determine the distance and direction to a specific feature and linking to other apps for turn-by-turn walking or driving directions.
- **Utility networks:** Particularly valuable in the utility sector, Field Maps integrates data about essential infrastructure, such as pipes, wires, regulator stations, and substations. Features such as network tracing, viewing connectivity associations, and accessing container views are available, supporting critical field activities.
- **Location sharing:** This feature allows for the sharing of mobile workers' live locations, enhancing safety and collaboration by keeping both office staff and mobile workers informed of one another's whereabouts in real time.
- **Map markup:** Allows mobile workers to sketch and note on the map, for their own information, similar to drawing on a paper map. they can also share their markups with others, including colleagues in the field.

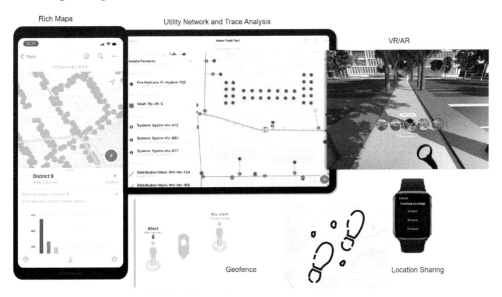

Figure 2.1. ArcGIS Field Maps provides rich maps with diverse data content, including utility networks and analysis tools, such as tracing, location sharing, geofencing, and VR/AR capabilities for situational awareness indoors and outdoors.

Feature layer pop-ups versus forms

Feature layers can have pop-ups and forms. they serve the following different purposes.
- **Pop-ups:** Pop-ups are primarily for displaying information and for situational awareness. they are read-only—users cannot edit the information displayed within a pop-up. Pop-ups efficiently provide a quick overview of the attributes associated with a feature when clicked on the map. they can display a variety of content, including attribute fields, Arcade expressions based on those fields, attachments, and charts, making them a versatile tool for information delivery without interaction capabilities.
- **Forms:** Field Maps forms are primarily designed for data input and editing and can also display read-only information, making them a dual-purpose tool that supports data viewing and editing. Forms offer customization options tailored toward data collection needs. they can be designed to guide the user through a workflow, ensuring data consistency and completeness. Forms support various input types, such as text, numeric, date, and choice lists, and can include conditional visibility of fields based on previous entries. This makes them effective for structured data entry in scenarios such as inspections or surveys.

Feature layer styles versus feature templates

Layer styles and feature templates serve distinct but complementary roles in the visualization of and interaction with geographic data.
- **Layer styles:** Layer styles in ArcGIS focus on visualization. they dictate how features within a layer are displayed on the map. Styles can be adjusted based on the attributes of the data to represent different variables, such as size, color, and symbol type. This aids in visually communicating information. these styles are applied solely for viewing purposes, making it easier for users to discern trends, patterns, and outliers in the data based on visual cues.
- **Feature templates:** Feature templates, on the other hand, are used during the editing mode. they are preconfigured setups for creating new features within a layer. these templates include preset attributes and default values that expedite the data entry process, ensure consistency, and help maintain data integrity during the creation of new records. they define what a new feature will look like and what attribute values it will initially carry when added to the map.

Although layer styles and feature templates often serve different phases of data handling—viewing and editing, respectively—it's generally beneficial for them to be consistent with each other. Consistency ensures that users do not experience confusion or a cognitive disconnect between how data is entered and how it is displayed. However, depending on specific use cases, they can also be deliberately varied. For example, an editable feature might be styled differently to highlight that it is under review or has been recently updated.

Geofences

Geofencing uses location-based services to create virtual boundaries around designated geographic areas, playing a crucial role in managing location sharing capabilities. Within Field Maps Designer, map authors can establish Geofences using polygon layers or buffers around various geometries. they can configure specific alert messages for when users enter or exit these Geofences and, crucially, control the automatic enabling or disabling of location sharing, as depicted in figure 2.2. When mobile workers using the Field Maps app cross these boundaries, the system not only alerts them but can also turn their location sharing status on or off automatically. This capability ensures that location sharing is managed precisely, safeguarding privacy while maintaining operational efficiency. By implementing Geofences effectively, organizations can bolster safety, ensure compliance, streamline data collection, and optimize resource allocation. In essence, embracing geofencing technology enables field teams to adapt more responsively and operate more efficiently within the dynamic conditions of their environment.

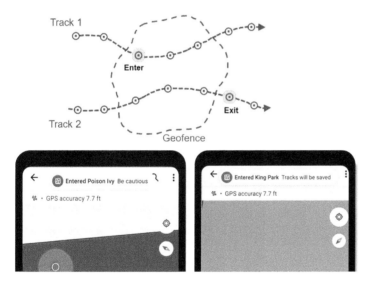

Figure 2.2. Geofences can automatically trigger alert messages and enable location sharing when mobile workers enter or exit the defined areas.

Data security and views

Data security is crucial for Mobile GIS workflows, especially when it involves granting data reading and editing access to the public. You can manage this access through a combination of hosted feature layer views, editing settings, and sharing properties. Views enable different groups of people to have varying levels of access to the same data. For mobile workers, and particularly the public, you might create a view that allows editors to see and edit only their own data or to add data without being able to see anything, including what they have added. For reviewers, you typically need to create a view that allows them to see and occasionally edit all the data contributed by all users. This approach ensures that data security is maintained while allowing the necessary access for different user roles (figure 2.3).

Related tables are common in Mobile GIS workflows. For example, you might need to inspect fire hydrants quarterly or maintain the status log of a service request. these scenarios require a layer with a related table to store the one-to-many relationship, making your database schema more scalable.

Figure 2.3. A view and its source feature layer reference the same data. You can independently set how the view is shared with others, how it's drawn, what features are displayed (filtering), and whether the layer can be edited.

Tutorial 2: Design for situational awareness and inspections with related records

In this tutorial, you will design a Mobile GIS solution to bring hydrant data to mobile workers for understanding the existing hydrants and their past inspections and for them to collect new hydrants and perform new inspections.

Data: You are provided with a zip file of a hydrant file geodatabase. It has a layer of existing hydrants and a related table of their previous inspections. To streamline and accelerate the tutorial, the layer and table contain a limited number of records. This setup allows for quicker demonstrations and easier learning.

Requirements:
- **Situational awareness capabilities**
  - **Hydrant visualization:** Implement visually distinct symbols for existing hydrants that are easily recognizable outdoors, with interactive pop-ups providing essential hydrant information quickly.
  - **Historical inspection data access:** Enable mobile workers to directly access and review historical inspection records associated with each hydrant, facilitating immediate understanding of past issues and maintenance.
  - **Geospatial context:** Incorporate additional layers that may influence hydrant inspection activities.
  - **Geofencing and alerts:** Set up geofencing to provide alerts to mobile workers about specific conditions when entering or exiting designated areas, enhancing situational responsiveness and safety.
- **Data collection and inspection capabilities**
  - Adding new hydrants: Allow mobile workers to add new hydrants into the system, capturing essential attributes and media attachments.
  - **Access control:** Allow mobile workers to add hydrants and inspections but prevent them from deleting any records.
  - Performing inspections: Use smart forms for inspections on existing and new hydrants. these forms should feature conditional visibility and automatic calculation expressions to streamline the inspection process and improve data accuracy.
  - **Related records management:** Link new inspections seamlessly to the corresponding hydrant records.

2.1: Create a hosted feature layer from an existing dataset

GIS professionals commonly use ArcGIS Pro to design and construct geodatabases ranging from basic to complex, including enterprise and file geodatabases, and publish them as feature layers for use with Web GIS and Mobile GIS. In this section, you will download a provided file geodatabase that contains existing hydrant data and publish it to ArcGIS Online as a hosted feature layer.

1. Start a web browser and navigate to https://arcg.is/1mzGLS1. Click Download.

2. Navigate to ArcGIS Online (www.arcgis.com or the URL of your ArcGIS Online organization) and sign in.

3. Click Content and then click New Item.

   The new item window opens. It has an upload area for you to upload files.

4. In the New Item window, click Your Device, browse to the zip file you downloaded, select it, and click Open.

5. For File Type, choose File Geodatabase. Click Next.

   The default option will add the zip file to ArcGIS Online and create a hosted feature layer.

6. For Title, type Fire Hydrants. You may append your name to the title if it's not available.

7. For Folder, choose Create New Folder, type Chapter2 as the folder name, press Enter, and click Save.

   After the data is published, the item details of the feature layer will appear.

   Next, you will review the data schema.

8. Observe that the feature layer has a sublayer named Hydrants and a subtable named Inspections.

9. Click the Data tab. Preview the data in the Hydrants layer. Click Fields to switch to the Fields view and review its attribute fields.

10. Click the Layer drop-down list, click the Inspections table, and review the attribute fields of the table. Switch to the Table view and note it has several records.

    Next, you will review the attachment settings on the layer and table.

11. Click the Overview tab. Click the Hydrants layer. On the right, under Attachments, notice it has attachments enabled.

12. In the upper-left area of the page, click the back arrow to go back to the parent layer, Fire Hydrants.

13. Click the Inspections table and notice it has attachments enabled, too.

14. Click the back arrow to go back to the parent layer.

    Next, you will configure the settings of the feature layer.

15. Click the Settings tab. In the Feature Layer (Hosted) section, under Editing, check Enable Editing, Keep Track of Who Edited the Data, and Enable Sync. Click Save.

    Next, you will create a view for the mobile workers to use and keep the source layer for yourself.

16. Click the Overview tab. Click Create View Layer and choose View Layer.

17. Click Next and then click Next again.

18. For Title, type Fire Hydrants (Field) and click Create.

    After the view is created, the item page displays.

    The view can have different editing settings from its parent or source. Next, you will enable editing but prevent editors from deleting data.

19. Click the Settings tab. In the Feature Layer (Hosted, View) section, under Editing, apply the following settings:
    a. Check Enable Editing.
    b. Check Enable Sync.
    c. Uncheck Delete.

20. Click Save.

2.2: Configure web map layer style and pop-ups

In this section, you will add the view layer to a web map and configure its styles and pop-ups for awareness in the field.

1. Click the Overview tab.

   If you have left the page, navigate to ArcGIS Online, sign in, click Content, find the Fire Hydrants field view that you created, and click it to open its item page for details.

2. Click Open in Map Viewer.

   The layer is added to a new map.

3. In the Layers list, click the Options button (three dots) of the Fire Hydrants (Field) – Hydrants layer, click Rename, and update the Title to Hydrants.

4. On the left toolbar, click Tables.

5. In the Table list, click the Options button of the Fire Hydrants (Field) – Inspections table, click Rename, and update the Title to Inspections.

6. On the left toolbar, click Layers and ensure that the Hydrants layer is selected.

7. On the right toolbar, click Styles.

   The layer is currently displayed using a single symbol. Next, you will change its style to a hydrant icon.

8. Under Location (Single Symbol), click Style Options.

9. In the Styles pane, under Symbol Style, click the Edit button.

10. In the Symbol Style pane, apply the following settings:
    a. Click Basic Point.
    b. Click the Category drop-down list and choose Government.
    c. Click List View and search for Hydrant.
    d. Choose Hydrants-Inspected (red hydrant button).

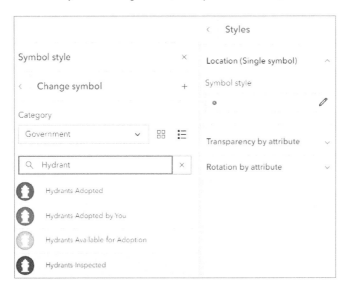

Chapter 2: Situational awareness and one-to-many inspections using Field Maps   47

    e. Click Done.
    f. Set the size to 20 px.
    g. Click Done and then click Done again to close the Styles pane.
    h. Click Save and Open and then click Save As to save your map. Set the title as Hydrant Inspection and set the folder as Chapter2.

The hydrants with the IDs of Esri-1 and CP-1 have attached photos and associated inspections. Next, you will find one of them and preview its pop-up.

11. In the Layers list, click the Options button of the Hydrants layer and click Show Table.

12. In the Hydrants table, scroll to the right of the table and click the Asset ID field.

    The values are sorted in ascending order.

13. In the table, find the Asset ID of CP-1 or Esri-1. Check the box of the record to select it.

14. Above the Home button, click the Zoom to Selection button to zoom to the selected hydrant.

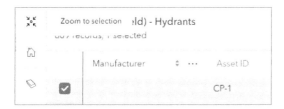

The hydrant is highlighted on the map.

15. Click the hydrant, review the default pop-up, and leave the pop-up open.

    As you configure the pop-up in the following steps, it will update immediately for you to preview.

16. On the right toolbar, click Pop-ups.

17. In the Pop-ups pane, apply the following settings:
    a. Click Title and set the title as Hydrant: , with a space after the colon.
    b. Next to the Title text box, click Add Field { } and choose the Asset ID ({assetid}) field.

Now the title looks like `Hydrant: {assetid}`.

Next, you will select the fields to display in the pop-up content section.

18. Click Fields List.

19. Click Select Fields.

20. Check the Outlet Size, Manufacturer, and Install Date fields and uncheck all other fields. Drag the fields to arrange in the order you prefer.

    Next, you will add attachments to the pop-up.

21. If the Attachments feature is not already selected, at the bottom of the Pop-ups pane, click Add Content. Click Attachments.

    Next, you will add links to previous inspections.

22. Click Add Content.

23. Click Related Records and apply the following settings:
    a. For Title, type Inspections.
    b. For Sort By, choose Creation Date.
    c. For Sort Order, choose Descending.
    d. Leave Preview Count at 1.

    Next, you will configure the pop-up of the table.

24. On the left toolbar, click Tables, and in the Tables list, click Inspections.

    The Inspections table is selected automatically.

25. On the right toolbar, click Pop-ups and apply the following settings:
    a. Set the title as {inspectiontype} on {insdate}.
    b. In the Fields list, check all the fields except Inspection Type and Inspection Date.
    c. If the Attachments feature is not already selected, click Add Content and click Attachments.

26. Click Save and Open and then click Save to save your map.

Chapter 2: Situational awareness and one-to-many inspections using Field Maps   49

2.3: Configure layer and related table forms

In this section, you'll configure the forms for both the layer and the related table for your mobile workers to collect new hydrants and add inspections.

1. In the upper right of the page, click the app launcher and click Field Maps Designer.

   If you have left the page of the previous section, navigate to ArcGIS Online. Sign in and click the app launcher.

2. Click the Hydrant Inspection map to open it.

   The Forms tab displays.

3. In the Forms pane, click Layers to expand it. Click the Hydrants layer.

   The blank form canvas displays.

4. In the Form builder pane, drag the Asset ID, Outlet Size, Manufacturer, and Installation Date fields to the canvas.

Next, you will configure the Asset ID field.

5. On the form, click the Asset ID field. In the Properties pane, apply the following settings:
   a. Under Logic, check Required.
   b. Next to Calculated Expression, click the Expressions button (gear).
   c. Click New Expression.
   d. In the Expression builder window, set the Title to Build Asset ID.
   e. In the script area, copy the following code from this page: links.esri.com/https://arcg.is/OC51v.

   ```
   var newId = "h-" + Number(Now())
   var username = GetUser(Portal('https://www.arcgis.com')).username;
   if (!IsEmpty(username)) {
     newId += "-" + username
   }
   return newId
   ```

   f. Click Run. Notice the output is like h-milliseconds-{your username}. This can make the asset ID unique.
   g. In the script area, replace the above code with the following lines of code. Copy them from this page: https://arcg.is/OC51v.

   ```
   if (IsEmpty($feature.assetid)) {
     var newId = "h-" + Number(Now())
     var username = GetUser(Portal('https://www.arcgis.com')).username
      if (!IsEmpty(username)) {
     newId += "-" + username
     }
     return newId
   } else {
     return $feature.assetid
   }
   ```

   h. Click Run and notice that the output is an existing ID.
   i. Click Done.

The new code will preserve the existing ID when editing a hydrant and will generate a new ID when adding a new hydrant.

Next, you will configure the installation date.

6. On the form, click the Installation Date field. In the Properties pane, apply the following settings:
   a. Check Required.
   b. Next to Calculated Expression, click the Expressions button.
   c. Click New Expression.
   d. In the Expression builder window, set the Title as Calc Installation Date.
   e. In the script area, copy the following code from this page: https://arcg.is/OC51v.
   ```
   if (IsEmpty($feature.installation_date)) {
     return Now()
   } else {
     return $feature.installation_date
   }
   ```
   f. Click Run and notice the output.
   g. Click Done.

   The code will preserve the existing installation dates and use the current date when the existing installation dates are null.

7. At the top of the page, next to the Undo/Redo buttons, click Save to Map to save the form.

   Next, you will configure the form for the table.

8. In the Forms pane, click Tables to expand it and click the Inspections table.

   The blank form canvas displays.

9. On the Form builder pane, scroll down the Fields list.

10. Press Ctrl and click the Inspection Type, Inspector, Inspection Date, and Notes fields to select them.

11. Drag them to the canvas.

12. Set Inspection Type as required.

13. Set Inspector Required and with the following calculation expression. Title the expression as Calc Full Name. Copy the script at https://arcg.is/OC51v.
    ```
    return GetUser(Portal('https://arcgis.com')).fullName
    ```

14. Set the Inspection Date required with the following calculation expression. Title the expression as Calc Now. Copy the script at https://arcg.is/OC51v.
    ```
    return Now()
    ```

Next, you will add a group for the Flow Test inspection type.

15. Close the Properties pane.

16. In the Form builder pane, expand the Form elements.

17. Under Layout, drag a Group element onto the form canvas.

18. In the Properties pane, set the group's Display name as Flow Test.

19. Under Logic, click the Expression button next to Visible and click New Expression.

20. In the Expression builder window, set Title as Is Flow Test. For the condition, set it as `Inspection Type is Flow Test`. Click Done.

21. In the Form builder pane, press Ctrl, select the following fields, and drag to the Flow Test group and change their properties:
    a. Is the Hydrant Operable? Make its Input type Switch.
    b. Pressure: Make it Required.
    c. Flow Rate: Set to Visible.

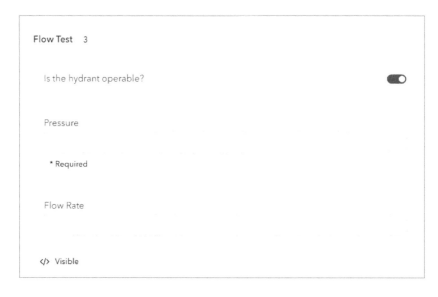

Next, you will add a group for the Maintenance Inspection type.

22. Close the Properties pane.

23. In the Form builder pane, expand the Form elements and drag a Group element onto the form canvas.

24. In the Properties pane, set the group's Display name as Maintenance Inspection.

25. Click the Expression button next to Visible and click New Expression.

26. In the Expression builder window, set Title as Is Maintenance Inspection. For the condition, set it to `Inspection Type is Maintenance`. Click Done.

27. In the Form builder pane, drag the following fields to the Maintenance Inspection group. Set their Input type to Switch and set them as Required:
    a. Is the Hydrant Leaking?
    b. Are Chains Missing?
    c. Needs Lubrication?
    d. Needs to Be Painted?

28. Click Save to Map to save the form.

2.4: Configure Geofences

In this section, you will enhance your web map by adding a provided Geofence layer, which will serve as a reference layer. Additionally, you will sketch a Geofence polygon and configure the associated Geofence messages to aid in field operations.

1. On the left toolbar, click Open > Map Viewer.

    If you closed your browser after the previous section, open a web browser and sign into ArcGIS Online. Click the app launcher, click Field Maps Designer, and then open the Hydrant Inspection web map.

    The web map opens and displays the Layers list.

2. In the Layers list, click Add.

3. Click My Content to open the drop-down list and click ArcGIS Online.

4. Search for owner:GTKMobileGIS Geofences.

5. In the results, find the Geofence Polygons layer and click Add to add it to your map.

    Next, you will reposition the layer below other layers and disable its pop-ups. This prevents it from obscuring other layers and avoids distracting mobile workers with pop-ups.

6. Next to Add Layer, click the back button.

7. In the Layers list, drag Geofence Polygons to a position beneath the Hydrants layer.

8. With Geofence Polygons selected, on the right toolbar, click Pop-ups.

9. Uncheck Enable Pop-ups.

10. Click Save and Open and then click Save to save the web map.

    Next, you will create a polygon to define a zone. The polygon should encompass your current location or be close enough for you to physically enter and exit the zone to trigger the messages in section 2.6.

11. On the right toolbar, click edit.

12. Under Create Feature, under Geofence Polygons, click New Feature.

13. Draw your zone as a polygon on the map.

    > **Note:** Geofence polygons are typically small and local. To minimize interference from other data, the filter on the Geofence Polygons view automatically excludes any polygons larger than 26 sq km, about 10 sq mi., or older than four months.

14. In the Create Features pane, type the zone name and messages for both entry and exit. For example:
    - Zone name: construction zone or school campus.
    - Message on Enter: Helmet required or No honking during class time.
    - Message on Exit: Helmet no longer required or Honking restriction lifted.

15. Click Create.

16. Close Map Viewer.

    The page will return to Field Maps Designer and reload your web map.

17. On the left toolbar, click Geofences.

18. Click Add Geofence.

    The New Geofence pane opens.

19. In the New Geofence pane, set the Name as Alerts on Zones.

20. For Layer, choose Geofence Polygons.

21. For Action Type, leave it as Location Alert.

22. For Message on Enter, type Entering. Click {} to select the zone_name field. Type a period and click {} to select the msg_on_enter field.

    Your text in the Message text area should read Entering {zone_name}. {msg_on_enter}.

23. Check On Exit.

24. Set the Message on Exit as Exiting {zone_name}. {msg_on_exit}.

25. Click Save.

The Geofence is created and saved to your map.

2.5: Configure web map bookmarks, filters, and search

Mobile workers frequently need to navigate to specific predefined areas, filter layers to display only certain hydrants based on attribute values, and search for specific hydrants. these functionalities can be configured using map filters, searches, and bookmarks to streamline their tasks.

First, you'll create a bookmark of Citrus Plaza in Redlands, California.

1. On the left toolbar of Field Maps Designer, click Open > Map Viewer.

2. Next to the right toolbar, click the Search button.

3. Search for Citrus Plaza, Redlands, CA, and press Enter.

4. On the left toolbar, click Bookmarks.

5. In the Bookmarks pane, click Add Bookmark, set the title as Citrus Plaza, Redlands, and click Add.

6. Zoom out to an extent that encompasses all the hydrants and your current location.

7. Click Save and Open and then click Save to save your web map.

   Next, you will configure the filters on the hydrants layer.

8. Close Map Viewer.

9. On the left toolbar of Field Maps Designer, click App Settings.

10. Click Layer Filters and apply the following settings:
    a. Click Selected Feature Layers and Fields.

Chapter 2: Situational awareness and one-to-many inspections using Field Maps   57

    b. Expand the Hydrants layer and check the Asset ID, Outlet Size, and Manufacturer fields.

       The Geofence layer cannot be filtered.

11. Click Save Changes.

Next, you will configure the thumbnail of the web map and the search settings on the hydrant layer.

12. On the left toolbar, click Open > Item Details.

13. Click edit Thumbnail.

14. Click Browse and select the hydrant image provided online.

You may download and use the sample hydrant thumbnail at https://arcg.is/1CSmzW0.

15. Click the Settings tab.

16. Scroll down to the Application Settings section and apply the following settings:
    a. Leave Enable Search checked.
    b. Check By Layer.
    c. Select Hydrants, Asset ID, and Contains.

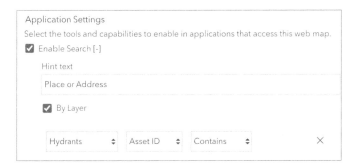

The setting will search for assets with IDs containing the letters you specify in section 2.6.

17. Click Save.

2.6: Collect related records using situational awareness

Typically, you should share the web map and layers with your mobile workers through a group to ensure that they can access and use your web map effectively. As the author of the web map and layers, you can always access the web map.

1. On your mobile device, find the Field Maps app and open it. Sign in with your ArcGIS Online account.

2. Find the Hydrant Inspection web map you created in the previous section and open it. If you haven't completed the web map, on your smart device, either scan the QR code below or go to https://arcg.is/LaLz1 to open the provided Hydrant Inspection web map.

When you open a map containing Geofences, you will be prompted to enable location sharing or location alerts, based on your configuration settings.

3. For the Location Alerts message: The map requires your location to alert you when you approach or leave specific areas. Tap Enable.

4. If prompted that Field Maps needs to access your motion and fitness activity, tap Allow.
    a. If you are within the Geofence polygon that you or others have defined, you should receive one or more entering alerts.
    b. If you are not within a Geofence, you can walk out to enter and exit a nearby Geofence you have defined or add a polygon to the Geofence layer using Field Maps directly.

Note: If you do not receive an alert from a Geofence you just added, which encloses your current location, tap the back button to return to the list of maps, tap the Options button next to the Hydrant Inspection map, and select Reload Map.

Note: If you are in a class, you may receive multiple alerts from nearby Geofence polygons created by other students. To minimize interference from others' data, the filter on the Geofence Polygons view automatically excludes any polygons larger than 26 sq km or older than four months.

Next, you will explore the overflow menu options including basemaps and bookmarks.

5. On the upper toolbar, tap the Options button to review the available menu options.

6. Tap Basemap, try several basemap options, and close the Basemap gallery.

7. Tap the Options button again, choose Bookmarks, and tap the Citrus Plaza bookmark to zoom to the extent of the bookmark.

Next, you will explore layer filters.

8. On the upper toolbar, tap the Layers button.

   The Layers list appears.

9. Tap the Hydrants layer to zoom out to the extent of the layer.

10. In the Map Layers section, tap the Filter button (three lines in a blue circle) next to the Hydrants layer.

11. Either tap Outlet Size and choose an option or tap Manufacturer and choose an option and tap Done.

The Hydrants layer now displays only those hydrants that match the filter criteria.

12. Tap the Filter button next to the Hydrants layer again, tap Reset, and tap Done.

    The filter is removed and all hydrants are displayed.

    Next, you will explore the web map search capabilities you configured in the previous section.

13. On the upper toolbar, tap the Search button.

14. Type CP-1 in the search bar and tap Search.

    Hydrants with asset IDs containing CP-1 display, including CP-1 and CP-18 and so on.

15. Tap the hydrant with the ID of CP-1.

    The map zooms to the hydrant and the pop-up you configured.

16. On the lower toolbar, tap the Actions button (three dots) and choose Compass. Observe the distance and bearing from your current location to the hydrant.

17. Tap the Actions button and choose Directions. Choose an app—for example, Google Maps—for directions to the hydrant.

18. Return to the Field Maps app.

    Next, you will review the pop-ups of the hydrant and the related table.

19. Swipe the hydrant pop-up to the top to view all details, including the hydrant's attributes, photo(s), and link to the past inspections.

20. Tap on a photo to review it and tap Done.

21. Tap Inspections to view the list of inspections.

    The inspection types and inspection dates are displayed. these are the pop-up titles of the table, as you configured in section 2.2.

22. Tap an inspection to view its details.

23. Tap Close to close the inspection details.

    Next, you will add a new inspection.

24. Tap Add in the Inspections list.

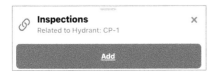

25. Experiment with the table form and test the conditional visibility and calculation expressions you configured.

    Your name and the inspection dates are filled in automatically because of the calculation expressions.

26. Complete the form and optionally attach a photo, and then tap Submit.

    Your new inspection is submitted. The inspection information is displayed immediately.

27. Tap the Search button and tap it again to clear the search.

    The link to inspections you just used is from the pop-up. Field Maps also provides access to related records through the Related Records button.

28. In the lower right of the screen, tap Add to add a new hydrant either at your current location or a different one.

29. Select the hydrant icon, fill in the hydrant form, optionally attach a photo, and tap Submit.

30. In the lower left, tap Related Records.

31. Tap Add, complete the Inspection form, and tap Submit.

In this tutorial, you published an existing dataset of hydrants to ArcGIS Online as a feature layer, configured the layer's style, set up pop-ups for both the layer and table, created bookmarks, layer filters, and searches, and established Geofences. You then explored the existing data and the capabilities from your settings in the Field Maps mobile app and gained situational awareness in the field.

Additionally, you developed a feature layer view specifically for mobile workers and improved data security by preventing them from accidental deletions of hydrants and past inspections. You applied and expanded on the form design and Arcade scripting skills acquired from the previous chapter. Using these forms, you collected data on hydrants and their related inspections using the field app. The one-to-many relationship between the hydrants and inspections exemplifies a typical inspection data schema and workflow applicable in various real-world scenarios.

Assignment 2: Use Field Maps for situational awareness and inspections with a one-to-many relationship

Requirements:
1. Use the sample data provided in this tutorial along with the feature layer you have published.
2. Add two new fields to the source hydrants layer and another two fields to the source inspections table.
3. Update the view using the Update View button on the view's Settings tab.
4. Enhance your web map by applying appropriate styling to both the hydrant and Geofence layers, differentiating them from those in the tutorial.
5. Configure the pop-ups to display attribute fields as text rather than a field list.
6. Incorporate the newly added fields into the forms for both the Hydrants layer and the inspection table.
7. Ensure that at least one of the new fields includes an Arcade calculation expression.
8. Share your layers and map either publicly or with your instructor through a group.

What to submit:
- The URL or QR code linking to your web map.

Chapter 3
Formcentric data collection using ArcGIS Survey123

Objectives
- Understand the components, capabilities, and workflow of Survey123.
- Create smart forms with calculations and dynamic lists using Web Designer.
- Understand the layers and views created by Survey123.
- Collect data using the Survey123 field app.
- Learn to use Survey123 Connect to implement advanced logic.

Introduction
In the previous two chapters, you learned about mapcentric Mobile GIS technology, specifically the use of Field Maps. This chapter introduces ArcGIS Survey123, a formcentric Mobile GIS technology. Survey123 bridges the gap between the extensive need for forms and their lack of geospatial characteristics, enabling users to create, share, and analyze location-aware surveys and mobile smart forms with ease. The chapter begins by introducing Survey123 components, including Web Designer, Survey123 Connect, the field app, and the website, followed by an explanation of the workflow and advancements in artificial intelligence (AI) integration. The tutorial sections will guide you through using Survey123 in various roles: as an author creating forms using Web Designer and adding advanced features using Survey123 Connect; as a mobile worker collecting data with the Survey123 field app; and as a GIS manager monitoring data collection progress and analyzing collected data.

The need for formcentric data collection
In today's business landscape, paper, PDF, and online forms are indispensable formats that organizations rely on for their daily operations. there had been a significant gap in the technology: the systematic neglect of location as a valuable data point. This oversight deprives organizations of critical opportunities to harness geography for enhancing data collection, visualization, and decision-making processes.

Survey123 fills the gap as a formcentric solution with sophisticated location intelligence capabilities. It allows users to easily create, share, and analyze location-aware surveys and mobile smart forms with skip logic (skipping questions based on the answers to earlier questions), defaults, validation, calculation, and more (figure 3.1). With the use of Survey123, users can design smart forms to collect research data, conduct inspections, assess damage, generate legal documents, and more. they can also visualize and analyze the results to support decision-making.

Figure 3.1. Survey123 field apps support formcentric data collection.

Survey123 workflow and components

The "1-2-3" in the product name reflects the philosophy of the product and workflow. Survey123 provides components to (1) create surveys, (2) gather responses, and (3) analyze the results. Survey123 (figure 3.2) offers the following components and capabilities.
- **Web Designer and Survey123 Connect**:
  - For designing intelligent surveys with a variety of geospatial and nongeospatial question types. The smart forms support coded values, defaults, conditional visibility of fields, conditional requirements for fields, field calculations based on other fields, and geoenrichment of data through querying or intersecting web layers dynamically.

- For designing engaging user interfaces with HTML and CSS; incorporate multiple media types, including audio, images, and videos; and use grid themes, picture sketches, and watermarks.
- **Survey123 field app and web app:** For data capture, using an intuitive, formcentric data-gathering user experience. The web app is web browser–based and works on all platforms. The field app is available for iOS, Android, and Windows platforms and supports offline mode.
- **Survey123 website:** Used to manage all surveys, view and analyze data, and generate feature reports.

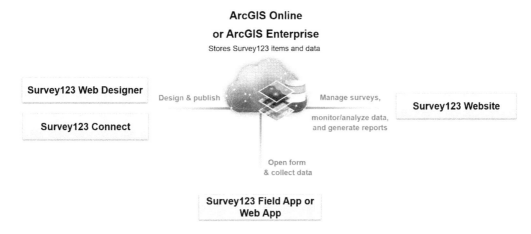

Figure 3.2. The components and workflow for using Survey123.

ArcGIS uses advancements in AI to enhance the experience of GIS users, accelerate decision-making, and streamline common GIS workflows. Survey123 is at the forefront of these advancements, incorporating key developments in GeoAI and AI Assistants. For a detailed exploration of these innovations, refer to chapter 9.

Survey123 Web Designer, Connect, and XLSForm basics

Both Survey123 Web Designer and Connect can design smart forms. Each possesses unique capabilities, and both are suited to different needs (figure 3.3).
- Survey123 Web Designer offers an easy-to-use interface, providing an entry-level experience in a What You See Is What You Get (WYSIWYG) environment. This allows for building smart forms quickly with drag-and-drop design capabilities and wizard-style settings for conditions and calculations, guiding authors through the creation of the most common types of survey questions and behaviors.

- Survey123 Connect provides survey authors the comprehensive XLSForm authoring experience, offering them full control over survey design and behavior. This environment is ideal for complex logic, an advanced user interface, connecting to existing feature services, and using JavaScript to accomplish extended capabilities.

XLSForm is an open standard that simplifies the authoring of forms. Survey123 Connect facilitates the design, preview, and publication of XLSForms, using Microsoft Excel workbooks for form design. A workbook typically includes the following main worksheets.
- **Survey:** This sheet contains the full list of questions, detailing how they will appear and behave in the form.
- **Choices:** This sheet includes options or coded values for single-select and multiple-select questions.
- **Settings:** This optional sheet can define the form title, submission URL, and other settings. The submission URL is particularly important if you want your survey responses to save to or load data from existing feature layers, including hosted and enterprise geodatabase based, instead of creating new hosted feature layers.

Form authors primarily work with the Survey sheet. On this sheet, each row typically represents one question. Note that each question has many columns, including the following commonly used ones.
- **Type:** Defines question types, including those you have used in Web Designer, such as select_one, image, text, and geopoint.
- **Name:** Defines the field name in the underlying feature layer.
- **Label:** Defines the field aliases and the label on the survey. Labels can be styled using basic HTML and CSS.
- **Required:** A value of yes means that the question must be answered.
- **Required_message:** Stores the message that users see if they do not fill out the required question.
- **Appearance:** Modifies how the question will appear or behave—for example, compact, autocomplete, and hidden.
- **Relevant:** Defines conditional visibility of questions.
- **Default:** Defines a formula that automatically populates the question when the survey is loaded.
- **Constraint:** Defines rules to restrict the acceptable inputs for a question.
- **Calculation:** Allows users to create formulas to perform integer, string, and geospatial-based calculations. For more details, refer to https://arcg.is/1jXzDv0.
- **bind::esri:fieldLength:** Specifies the field length in the feature layer schema.

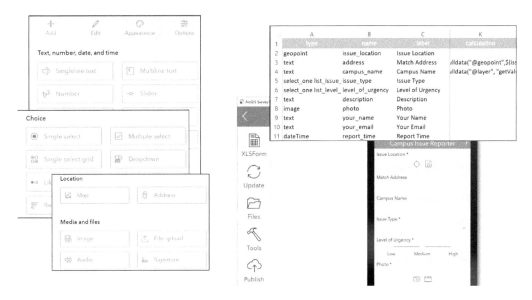

Figure 3.3. Survey123 Web Designer (*left*) and Survey123 Connect (*right*).

Tutorial 3: Design smart surveys and collect data using Survey123

An organization needs a mobile solution with a smart form for its staff and the public to report nonemergency issues on campus and requests for services. *Campus* here refers to the site of a university, a company, or any other type of organization.

Requirements:
- The form must enable users to report the location of the issue, the type of issue, the urgency level, and a description of the issue; upload photos; and give the user's full name and email address.
- The form should automatically calculate the address, based on the reported location.
- The list of issue types should be extensible and editable without the need to re-create, republish, and redownload the form.
- The form should allow users to sketch on photos to highlight the issues.
- The form should establish and manage a one-to-many relationship between an issue and its multiple processing statuses.

Data: The following ArcGIS Online content items are provided:
- A feature layer that stores the list of issue types.
- A helper survey that allows users to add additional issue types to the above feature layer.

- A campus web map with a campus polygon feature layer.
- An Experience Builder web app that allows users to add additional campus polygons and names.

System requirements:
- An ArcGIS Online creator or publisher account.
- A mobile device for installing Survey123 for use in section 3.4.
- A Windows computer for installing Survey123 Connect for use in sections 3.5 and 3.6.

3.1: Design a basic form using Survey123 Web Designer

1. In a web browser, go to survey123.arcgis.com and sign in.

   Alternatively, you can go to your ArcGIS Online organization, sign in, click the app launcher, and click Survey123.

2. Click New Survey.

3. Under Blank Survey, click Get Started.

4. Click the Edit Survey Info button (pencil) on the toolbar.

5. In the Edit Survey Info window, apply the following settings:
   a. For Name, type Campus Issue Reporter.
   b. For Tags, type GTKMobileGIS, <your organization name>.
   c. For Summary, type Report nonemergency issues on campus and request for services.
   d. Click OK.

   The survey info is for the content item as it appears in ArcGIS Online. Next, you will configure the title and description that will display on the survey form.

   Survey123 Web Designer displays an empty survey on the left and the Design pane with a list of available question types on the right.

6. On the survey, click the Survey Title Not Set header. On the Edit tab of the Design pane, change the survey header text to Campus Issue Reporter.

7. On the survey, click Description Content for the Survey. On the Edit tab of the Design pane, change the survey description text to Report non-emergency issues on campus and request services.

8. Click Save.

Save often to avoid losing your work.

Next, you will add a Map question to the survey for collecting issue locations.

9. In the Design pane, click the Add tab and click Map to add it to the survey.

10. On the Edit tab, apply the following settings:
    a. For Label, type Issue Location.
    b. For Drawing tools, choose Point. Note the line and polygon options available.
    c. For Map and Extent, expand the list and choose Imagery Hybrid. Zoom to the extent of your campus or organization. Press the Shift key and draw a rectangle to define the extent.
    d. For Default location, choose Use Device Location and Ask for Location When Answering This Question.
    e. For Validation, choose This Is a Required Question.

The map question will use the device's location by default and will allow users to manually modify the location in the Survey123 mobile app.

Next, you will add a question for users to specify the issue type. You can use the Single Select question type or the Dropdown question type. Here you will use the latter so that you can make the list dynamic in the next section.

11. In the Design pane, click the Add tab, and click Dropdown to add it to the survey.

12. On the Edit tab, apply the following settings:
    a. For Label, type Issue Type.
    b. Set the choices to AC, Lighting, Graffiti, Pothole, Trash, and Recyclables. You can add the choices one by one or use Edit > Batch Edit Choices. If you choose the former, click the Add (+) button to add more choices.
    c. Choose Allow Other.
    d. For Appearance, choose Autocomplete.
    e. For Validation, choose This Is a Required Question.

Next, you will add a question for users to specify the level of urgence of the issue.

13. In the Design pane, click the Add tab, and click Likert Scale to add it to the survey.

14. On the Edit tab, apply the following settings:
    a. For Label, type Level of Urgency.
    b. Set Maximum value to 3.
    c. Set the choices to Low, Medium, and High.
    d. For Validation, choose This Is a Required Question.

Next, you will add a text question for users to describe the issue and set the visibility of this question so that it appears only if the users have set a Level of Urgency to Medium or High.

15. In the Design pane, click the Add tab and click Multiline Text to add it to the end of the survey.

16. On the Edit tab, apply the following settings:
    a. For Label, type Description.
    b. For Behavior, choose Visible.
    c. Under Visible, click Set Rule.
    d. Set the rule as Level of Urgency Is Not Low.

17. Click OK.

Next, you will add an Image question to allow users to upload a photo of the issue.

18. In the Design pane, click the Add tab and click Image to add it to the survey.

19. On the Edit tab, for Label, type **Photo**. Set it as a required question and set the minimum file count to 1.

Next, you will add questions to ask for the user's name and email.

20. In the Design pane, click the Add tab and click Singleline Text to add it to the end of the survey.

21. On the Edit tab, for Label, type **Your Name**. Set it as a required question and choose Cache Answer.

> **Note:** Cached answers are automatically populated the next time a survey is started, so users won't have to type their names each time they complete a survey.

22. In the Design pane, click the Add tab and click Email to add it to the survey.

23. On the Edit tab, for Label, type **Your Email**. Set it as a required question and choose Cache Answer.

Next, you will add a hidden question to automatically log the submission date and time without user input.

24. In the Design pane, click the Add tab and click Date and Time (choose the combined option) to add it to the survey.

25. On the Edit tab, configure the following settings:
    a. Label: Report Time.
    b. Default Value: Submitting date and time.
    c. Validation: This is a required question.
    d. Behavior: Uncheck Visible.

26. Click Save.

You have created a basic survey. You will further enhance the survey in the next section. Keep this page open for use in the next section.

3.2: Configure calculations and dynamic lists in Web Designer

In this section, you will configure the form to automatically do the following:
- Calculate the matching address of the issue location so that your end users won't have to type it manually and so that the service dispatchers and responders can understand the issue locations easier.
- Calculate the campus name where the issue location falls, which helps dispatchers discern whether the location is inside or outside the campus.
- Retrieve the issue types from a feature layer, allowing your organization to add, edit, and delete issue types without needing to modify and republish the survey and without requiring mobile users to redownload the survey.

> **Note:** If you have closed the survey, go to survey123.arcgis.com and sign in. Go to My Surveys, find the survey, and click its design button to open it in Web Designer.

1. In the survey, click the Issue Location question.

2. In the Design pane, click the Add tab. Click the Singleline Text button to add it to the survey.

   If the question is not placed under the Issue Location question, drag it to move it.

3. On the Edit tab, for Label, type Match Address.

4. Under Calculation, click Edit.

5. In the Edit Calculation pane, on the Extract Data tab, apply the following settings:
    a. Source: Confirm that Question is selected.
    b. Extract Property from Question: Issue Location.
    c. Select a Property: Match Address.

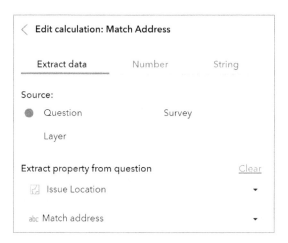

6. Click OK.

Next, you will add a question under Match Address to calculate the name of the campus where the issue location falls.

7. In the Design pane, click the Add tab and click Singleline Text to add it to the survey.

8. On the Edit tab, for Label, type Campus Name.

9. Under Calculation, click Edit.

10. In the Edit Calculation pane, on the Extract Data tab, apply the following settings:
    a. Source: Layer.
    b. Under Select Layer or Table, click Add.
    c. Click My Content to expand the options and choose ArcGIS Online.
    d. Search for campuses layer owner:GTKMobileGIS.
    e. Select the resulting Campuses layer.
    f. Click OK.
    g. Select Output Field: Name.
    h. Under Set Filter, turn on Extract Data by Location and choose Issue Location.

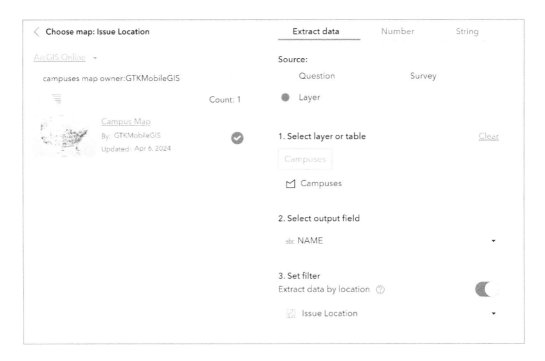

11. Click OK.

Typically, you wouldn't need to show the calculated Match Address and Campus Name to your users. Here, you leave them visible so that you can see the results when testing the form.

Next, you will make the Issue Type list dynamic.

12. On the survey form, click the Issue Type question.

13. On the Edit tab, under Choices, click Edit and apply the following settings:
    a. Click the From Layer tab.
    b. For Select a Feature Layer or Table, click Add.
    c. Click My Content to expand the options and choose ArcGIS Online.
    d. Search for Issue Types from readers owner:GTKMobileGIS. Select the result.
    e. Click OK.
    f. Select Choice Label Field: Issue Type.

14. Click OK.

    When you test the survey in section 3.4, you can add your own issue types using https://arcg.is/1ir8C10 and then see your issue type appear in the drop-down list.

    > Note: User-added issue types will be automatically removed after three weeks to keep the choice list clean.

    Next, you will change the map for the issue location question from a basemap to a web map with a relevant layer—the campuses layer. This layer will help your end users visually understand whether the location is inside or outside a campus boundary.

15. Click the Issue Location question.

16. Under Map and Extent, click Edit.

17. Click Organization's Basemap Gallery to expand the options and choose ArcGIS Online.

18. Search for Campus Map owner:GTKMobileGIS.

19. Select Campus Map in the search results.

20. Click OK.

    Next, you will review the theme and the thank-you screen.

21. Click the Appearance tab and then click various themes to try different options. Select a theme you like. Note the colors, text, and background image that you can edit.

22. Click the Options tab. Change the default message to Thank you! The issue you reported was submitted successfully.

23. Click Save.

You have enhanced your survey. You will publish it in the next section.

3.3: Publish your survey and review the items created

In this section, you will first preview the survey's user interface and behavior, examine the schema of the feature layer to be created, publish the survey, and then review the items created as well as the importance of feature views for data security.

> Note: If you have closed the survey, go to survey123.arcgis.com and sign in. Go to My Surveys, find the survey, and click its Design button to open it in Web Designer.

1. Click Preview.

2. Click on the map to define an issue location and observe the results of the address calculation. If prompted, adjust your browser settings to allow location services.

3. Click the Issue Type drop-down arrow and choose one.

4. Change the level of urgency to see how the Description question hides and reappears according to the visibility rule you previously configured.

5. Click Submit and note the messages regarding incomplete required fields.

6. Click the Phone, Tablet, and Desktop buttons to see how your survey responds to different screen sizes.

7. Click Close Preview.

8. Click Publish.

9. In the Publish Survey window, click Modify Schema.

10. Review the field names, types, lengths, and available choices.

11. Keep the schema as is but understand you can change the field names and lengths if needed.

    Publishing the survey form using Web Designer will create a new feature layer with the geometry type, attribute fields, and coded values you defined in the previous sections.

12. Click Publish.

13. When you see a window saying your survey has been published, click OK to continue.

    Next, you will share your survey with your field users, which can include all staff and the public.

14. On the menu bar, click Collaborate.

15. Click the Share Results tab. Share your survey with Everyone (public) and click Save.

    The publishing process creates a folder in your content, with the survey name as the folder name.

16. In your web browser, go to arcgis.com and sign in.

17. Click Content. Under My Content, on the left, find the folder Survey–Campus Issue Reporter and click it.

    The folder contains a form, a feature layer, and two views of the layer.
    - The Campus Issue Reporter layer is the source and is for you, the owner, to manage.
    - The Campus Issue Reporter_results layer is a hosted feature view and is reserved for sharing the survey results with the group of stakeholders who can see all records.
    - The Campus Issue Reporter_form layer is a view and is used in the survey form.

    Next, you will examine the settings on the Reporter_form view.

18. Click the Campus Issue Reporter_form View to open its item page.

19. Click Share. Notice that it is shared with the public.

20. Click the Settings tab. Scroll down and note the following settings:
    a. It is editing enabled.
    b. Users can only add records.
    c. Editors can see only their own records.

These view and edit settings ensure that end users can report only new issues, cannot edit or delete the issues they reported, and cannot see or edit issues reported by other users. This setup ensures both their access and data security simultaneously.

Your survey form is ready for your end users to use.

3.4: Collect data using the Survey123 field app and review data using the Survey123 website

Survey123 has editions for different platforms. This section uses the iOS native edition. other editions are similar.

> **Note:** If you haven't installed Survey123 on your mobile device, install it. Search for Survey123 in the App Store on your iPhone, Google Play on your Android phone, or the Microsoft Store on your Windows computer.

1. On your computer, in a web browser, go to survey123.arcgis.com and sign in.

2. Under My Surveys, find Campus Issue Reporter and click Collaborate (World).

3. On the Collaborate tab of your survey, choose Ask the User How to Open the Survey, in a Browser or in the Survey123 Field App.

4. Under Link, click the Show QR Code button to bring up the QR code.

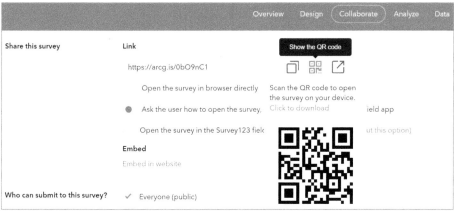

5. On your mobile device, open the Camera app and scan the QR code.

6. When prompted to Open in Browser or Open in the Survey123 Field App, choose the latter.

7. If prompted to sign in, do so and continue.

   Your survey will be downloaded and loaded in the Survey123 mobile app.

   Next, you will use the survey to collect data.

8. Tap Collect to start the survey.

9. For Issue Location, your current location can be used. You can also tap, zoom, and pan the map to choose a different location and then return to the form.

   The next question, Match Address, is calculated automatically, and the following question, Campus Name, may also be calculated.

   For locations without precise addresses, city or other place-names may be used. If locations do not overlap with campuses, the resulting campus name will be empty. If your organization's campus is not included in the campus layer, you can add a campus polygon by visiting https://arcg.is/1i0b8H0. Once it is added, return to the survey, set your location, and notice the updated Campus Name calculation.

10. Choose an Issue Type.

    As mentioned in section 3.2, you can add issue types using https://arcg.is/1ir8C10 and then reopen the survey to see your issue type appearing in the drop-down list.

11. Change the Level of Urgency. The Description question disappears if the Level of Urgency is set to Low but is shown if the level is Medium or High. Choose a final Level of Urgency and specify a description, if needed.

12. For Photo, tap the Camera button to take a photo or tap the Folder button to select a photo on your device.

13. Type your name and email address.

14. Tap the Submit button (check mark). You have the options to send it now, continue the survey, or save it. Tap Send Now.

    Next, you will review the data you've collected on the Survey123 website.

15. On your computer, return to the Survey123 website. On the navigation bar, click My Surveys, and find and click the Campus Issue Reporter survey.

    On the Overview tab, the total records collected, the total participants, and the time the records were collected are indicated.

16. Click the Analyze tab. Review the charts and summaries of Issue Types, submitters' names, and report time.

17. Click the Data tab and review the locations of the records on the map and the attributes in the table. Click a record on the map or in the table. The record is highlighted, and the details and photo can be reviewed.

18. Click Export and review the formats to which you can export the data.

19. Click Report and review the options to generate reports.

> Note: If the Report button is dimmed, it means you do not have the privilege to generate reports. You should contact your GIS administrator for assistance. Chapter 9 will teach you how to create feature report templates and generate reports automatically using webhooks.

3.5: Get started with Survey123 Connect (optional)

In this section, you will learn Survey123 Connect and make a couple of enhancements to the survey.

> Note: This section requires a Windows computer. On your Windows computer, if you haven't installed Survey123 Connect, download and install the app from the Microsoft Store (apps.microsoft.com/).

1. Start Survey123 Connect and sign in with your ArcGIS Online account.

   Survey123 Connect should list surveys you've created before.

2. You can start with the Campus Issue Reporter survey you created in section 3.2, or you can use an Excel file if you didn't complete that section.
    a. If you choose the former, find your Campus Issue Reporter survey in the list and download it. Click OK after the download completes. Then click the menu button (three dots) in the survey thumbnail and click Save As. Title it Campus Issue Reporter (connect). Click Create Survey and then open it.
    b. If you choose the latter, download the Excel file at https://arcg.is/15LjTS2 and drag it to Survey Connect to create the same survey as the one in section 3.2. Click OK after the survey is created.

   The survey form displays in preview mode.

   The survey created in section 3.2 uses a web map—the campus map. The web map is considered linked content, which requires that the new form you just created be published first for the web map to be usable in the survey.

3. Click Publish and then click Publish Survey. When publishing completes, click OK.

4. Click Linked Content, and then at the top of the window, click Link Content.

5. Click Online Map and search for Campus Map Link owner:GTKMobileGIS. In Filters, ensure that Only Search in <your organization name> is turned off.

6. Click Campus Map and click OK.

   The campus map used by the Issue Location question is now linked.

   Next, you will review the Excel workbook and learn the syntax of Survey123 Connect.

7. On the left, click the XLSForm button to open the Excel workbook.

8. Review the Excel workbook. At the bottom, notice the Survey, Choices, and Settings tabs.

9. On the survey sheet, review the columns for the questions. Understand the purpose of these columns: Type, Name, Label, Appearance, Required, Default, Relevant, Calculation, Constraints, and bind::esri:fieldLength.

10. Under the Calculation column, examine the syntax of the pulldata calculation formulas for the Match Address and Campus Name questions.
    - `pulldata("@geopoint",${issue_location},"reversegeocode.address.Match_addr")`
    - `pulldata("@layer", "getValueAt", "attributes.NAME", "https://services2.arcgis.com/j80Jz20at6Bi0thr/arcgis/rest/services/Colleges_and_Universities_Campuses/FeatureServer/0", ${issue_location}, "")`

    The reference page for the pulldata function is https://bit.ly/3VQhkv5.

11. Under the Appearance column, examine the syntax for the Issue Type question.

    `autocomplete search("list_issue_type?url=https://services2.arcgis.com/j80Jz20at6Bi0thr/arcgis/rest/services/survey123_f8fc446745a247299f3beea841c410b2_results/FeatureServer/0")`

    The reference page for the search appearance is https://bit.ly/3TSrJnd.

12. Under the Relevant column, examine the formula for the Description question.

```
(string-length(${level_of_urgency})>0)
and(${level_of_urgency}!='Low')
```

${question_name} refers to the value of a question. The above formula means that the Description question is visible when the level_of_urgency question has a value but the value is not low.

Next, you will make the Description question always visible, but it will be required only when the level of urgency is medium or high.

13. In the Excel worksheet, locate the row for the Description question, cut the formula in the Relevant column, and paste it to the Required_message column.

14. Save the Excel file and notice that the form preview updates to reflect the change.

15. In the form preview, change the Level of Urgency. Notice that the Description question remains always visible and becomes required only when the level of urgency is not low.

Next, you will change the appearance of the Photo question to allow users to sketch on the images so that they can highlight the issues on the photos.

16. In the Excel worksheet, find the appearance cell of the Photo question and type annotate.

The appearance now is `multiline annotate`.

Next, you will change the calculation mode of the Campus Name to `always`. This will ensure that this calculation is always executed whenever the issue location is added or moved.

17. In the Excel worksheet, find the bind::esri:parameters column of the Campus Name question and type calculationMode=always.

18. Click Publish to publish the survey.

19. Optionally, you can test the survey on your mobile device.

3.6: Add repeats and related tables using Survey123 Connect (optional)

A single request can have multiple statuses, such as received, assigned, and complete, among others. This necessitates a one-to-many relationship with a table. In this section, you will implement this requirement.

1. Start Survey123 Connect and sign in.

2. You can begin with the Campus Issue Reporter survey you created in section 3.5 or use a provided Excel file if you didn't complete section 3.5.
   - If you choose the former, locate the Campus Issue Reporter survey in your list of surveys. Click the menu button, click Save As, and rename it Campus Issue Reporter (repeat). Then click the new survey to open it.
   - If you choose the latter, download the Excel file from https://arcg.is/1WKvn90 and drag it into Survey123 Connect to create a survey identical to the one in section 3.5. Click OK after the survey is created.

   The new survey will display in preview mode.

3. Click Publish, click Publish It, and click OK when publishing completes.

   A newly published survey doesn't have any linked content. Next, you will link it to the campus web map, similar to the steps in the previous section.

4. Click Linked Content and then click Link Content.

5. Click Online Map and search for Campus Map Link owner:GTKMobileGIS. In Filters, turn off the option to Only Search in <your organization name>.

6. Click Campus Map and click OK.

   Next, you will add the list of choices to the Choices sheet for use by the Status question you will be adding.

7. In the Excel workbook, click the Choices sheet.

8. At the last row, add the following rows. Note list_status goes to the list_name column, and the rest of the two fields go to the Name and Label columns. To minimize typing, copy the rows from the Choices tab of this Excel file at https://arcg.is/1K8uaq1.

list_name	Name	label
list_status	Received	Received
list_status	Assigned	Assigned
list_status	In Progress	In Progress
list_status	Completed	Completed

Next, you will add the repeat and several questions to the Survey worksheet.

9. Click the Survey worksheet.

10. For the last row, add the rows as illustrated. To minimize typing, copy the rows from the Survey tab of the same Excel file in the previous step.

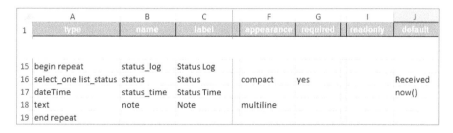

11. Save your Excel file.

The form preview will update. Next, you will test the repeat in Survey123 Connect. Alternatively, you can test the survey by scanning the QR code at https://arcg.is/1uKvHO1.

12. On the form preview, scroll down to see the repeat.

13. Choose a status—for example, Received.

14. Click the Add button to add another related record.

    The counter changes to 2 of 2. By using the repeat, you can add multiple related status records to the parent issue.

    For reporting issues, end users don't have to see the repeat and know the complex schema. To simplify the form for public users, you can hide the repeat by setting the appearance of the repeat to hidden. For dispatchers and responders, you can leave the repeat visible so that they can view all related statuses and add new ones.

15. Publish your survey.

    Next, you will examine the schema of the layer and the related table.

16. On the left toolbar, click More Actions and click View Item in your ArcGIS Online organization.

    The item page opens showing the details of your survey.

17. Under Layers, click the layer Campus Issue Reporter (repeat).

The item page of your feature layer opens.

The feature layer has a layer and a table named status_log.

18. Click the status_log table, click the Data tab, and click Field.

   The table has a relationship and a field named parentglobalid. This field is the foreign key related to the globalid field of the issues layer.

19. Optionally, you may open and test the survey on your mobile device.

   In this tutorial, you developed a mobile solution that enables both staff and the public to report nonemergency issues on campus and request services. The process began with Web Designer to construct a basic form, which was then enhanced to automatically identify and fill in the corresponding address and intersecting campus name. You made the issue list dynamic, which allows for updates independently without the need to re-create, republish, or redownload the form.

   Upon publishing the survey, you examined the survey itself, along with the associated feature layer and views created for managing data access and security. You practiced data collection using the Survey123 mobile app and reviewed the results on the Survey123 website. Later, you explored Survey123 Connect to further refine the form, including the addition of a repeat section to manage a one-to-many relationship between an issue and its statuses through a related table.

   This tutorial guided you through the form-based workflow, showcasing the versatility and capabilities of the Survey123 suite, including form design, data schema design, form publication, field data capture, and result review.

Assignment 3: Create a smart form using Survey123 for data collection

Create a smart form using Survey123 for one of the following scenarios:
- A law enforcement agency seeks to encourage citizens to report suspicious activities.
- A plant health inspection organization aims to motivate the public to report trees infested with insects.
- A public health department wants to enable citizens to report COVID-19 symptoms.
- Other topics distinct from the tutorial in this chapter.

Requirements:
Your survey must enable users to do the following:
- Specify locations on a map.
- Submit photos as evidence.
- Describe their observations or experiences.
- Choose one or multiple options, such as the type of suspicious activity, type of insects, type of symptoms, and so on.
- Provide their email and name.

Additionally, your survey must include the following:
- At least one conditional visibility rule.
- A calculation based on feature layers/services. You may use feature services from ArcGIS Living Atlas of the World or your own layers.

Your survey must be shared with the public or a group to which your instructor belongs.

What to submit:
- The URL to open your survey in a browser or the Survey123 field app.

Chapter 4
Rapid data collection using ArcGIS QuickCapture

Objectives
- Understand the components, capabilities, and workflow of QuickCapture.
- Create layer schemas and configure layer styles for use in QuickCapture.
- Design QuickCapture projects for rapid data collection.
- Author Arcade scripts for automatic field value calculations.
- Use the QuickCapture field app to capture data.
- Create QuickCapture projects for inspections based on oriented imagery.

Introduction
The previous chapters introduced Field Maps and Survey123, which offer precise control over map viewing and complex form handling. However, in some scenarios, data collection must be executed swiftly and without interruption, such as while riding or driving. In these cases, users cannot afford to stop and focus on placing locations on the map or filling in lengthy forms. Users require a method to collect data quickly and with minimal device interaction to ensure safety. This is where tools specialized for data collection at speed are required. This chapter will introduce QuickCapture, discussing its benefits, capabilities, and workflows, as well as its integration with drone technologies and oriented imagery. The tutorial will teach how to create layers by copying and registering existing feature layers, design layer styles for use in QuickCapture, configure projects using QuickCapture Designer, use device variables, perform Arcade calculations, and collect data using the QuickCapture field app. Furthermore, the tutorial will walk users through on creating a project for oriented imagery–based inspections using a QuickCapture template.

The need for data collection at speed using QuickCapture

In today's fast-paced world, the ability to capture real-time, location-specific data quickly and accurately is increasingly critical across various fields, such as the following.

- **Public safety:** During rapid windshield damage assessments, there is no time for extensive training. Responders need to quickly assess and record data on the go.
- **Utilities:** For pipeline inspections, it's essential to record maintenance issues and encroachments accurately, whether from the air or on foot, to ensure infrastructure integrity and compliance.
- **Conservation:** In invasive species mapping, conservationists often need to capture data while actively managing the environment, such as during spraying. A simple tap on a button allows them to begin capturing data lines without interruption.
- **Agriculture:** In high-stakes scenarios such as vineyard mapping for premium wine production, precision GPS and mobile GIS apps are essential to map hundreds of thousands of plants, ensuring precise management and cultivation of crops.
- **Law enforcement:** Officers need to record incidents and activities on the move, requiring a method that allows them to quickly document while patrolling.

QuickCapture is designed to meet such needs, providing a solution for users to capture data simply and at speed, whether from a moving vehicle, helicopter, bicycle, or all-terrain vehicle. It has two main components.

- **QuickCapture field app:** This app is designed for mobile workers to capture data rapidly. Its user interface (figure 4.1) features a series of large buttons, allowing users to quickly capture location, attributes, and photos with minimal attention required. The minimalistic interface enables users to conduct inspections by capturing locations, field conditions, and photos swiftly, without interrupting their primary tasks. With external GNSS receivers, QuickCapture can obtain high spatial accuracy. It can send the data back to the office for real-time analysis, eliminating the time spent manually processing handwritten notes.

Chapter 4: Rapid data collection using ArcGIS QuickCapture   91

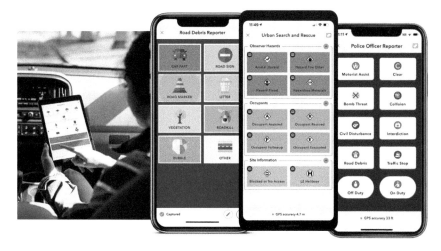

Figure 4.1. The QuickCapture field app features large, easy-to-use buttons that mask complex functionalities, enabling rapid data collection by mobile workers even while on the move.

- **QuickCapture Web Designer:** Web Designer provides a no-code approach for content creators to design QuickCapture projects. Web Designer (figure 4.2) allows you to create projects from feature layers or a curated collection of templates or by copying existing projects. You can configure the buttons and group appearances, location source, photos, videos, button user inputs, project user inputs, Arcade scripts, and webhooks for efficient data collection.

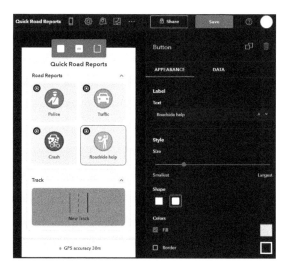

Figure 4.2. QuickCapture Web Designer provides a no-code approach for designing the look and feel of the user interface, configuring user interactions, and defining the logic behind the scenes.

Compared with Field Maps and Survey123, QuickCapture does not support editing existing data nor does it allow extensive data to be brought into the field. However, QuickCapture excels in rapid data collection, as previously discussed. A distinctive advantage is its ability to collect multiple features simultaneously. Users can activate multiple buttons at once to continuously stream points, lines, and polygons, either within the same layer or across multiple layers. For instance, while streaming a trail line, you can simultaneously press buttons to collect points of interest along the line. If you need to prevent simultaneous data capture from multiple buttons, they can be placed into an exclusive group.

QuickCapture workflow

The essential steps for using QuickCapture include the following workflow.

Step 1: Create feature layers with the schema you need to store the data to be captured.

QuickCapture requires at least one editable feature layer. You can create feature layers in many ways, as listed below. For security purposes, you often need to create multiple views for mobile workers, data reviewers, and a read-only audience.

- Build your feature layer and add fields interactively using ArcGIS Online or Portal for ArcGIS, as demonstrated in chapter 1.
- Copy existing feature layers.
- Create layers from feature layer templates.
- Create references to existing feature layers.
- Use Field Maps Web Designer.
- Use Survey123 Web Designer or Survey123 Connect.
- Upload and publish a zipped file geodatabase or shapefile.
- Use ArcGIS Pro to publish feature services, either hosted or based on an enterprise geodatabase.

Step 2: Style the feature layers for use in the QuickCapture project and project map.

In this optional step, you style a layer or create its feature templates and save the style or templates to the layer itself. When you add the layer to a QuickCapture project or other maps and apps, its style or templates will automatically be imported into the buttons, making its symbols consistent across many maps and apps.

Step 3: Create a new QuickCapture project with the layers.

- Create layer buttons from the layer symbols or change the buttons to use other symbols.
- Configure the locations, attributes, photos, videos, Arcade scripts, and webhooks.
- Preview, publish, and share your project.

Step 4: Collect data using the QuickCapture field app.
This step is typically performed by mobile workers.

Step 5: View the data collected in web maps or web apps.
This step can be performed in the office by data reviewers, decision-makers, and other stakeholders.

Integration with oriented imagery

Oriented imagery (OI) is an ArcGIS technology for managing, exploring, and viewing imagery captured from any angle. This technology supports images collected from aerial, drone, or terrestrial sensors, including oblique, bubble, 360-degree, street-side, and inspection imagery, among others. Integrating oriented imagery adds photographic context to your maps and geographic context to your images. New types of imagery, such as street-level images, provide valuable visual context to vector data representing on-the-ground assets and areas of interest. With oriented imagery, you can catalog, query, and visualize images in the context of a map to find and interpret the images you need, enhancing your understanding of their relationship to other geospatial data you manage.

The potential applications are wide-ranging:
- Use oriented images to provide situational awareness for first responders or mobile workers before they visit a site or asset.
- Manage and query an archive of images gathered over time with oriented imagery—for instance, when an issue such as vegetation encroachment is reported, quickly review all the images of that location to determine when it first occurred.
- Support inspection workflows by visualizing images of an asset from multiple directions.

QuickCapture Web Designer provides templates to quickly create projects supporting oriented imagery. You can also enable oriented imagery on a point layer yourself, adding fields to store metadata, such as the photo's location, heading, camera roll and pitch angles, and other optical characteristics. The QuickCapture field app automatically calculates and stores these metadata fields. Once oriented imagery is enabled for a project (figure 4.3), users will notice no difference in their user experience because everything is handled behind the scenes. Oriented imagery can be viewed using web apps built with the Oriented Imagery widget, Map Viewer in ArcGIS Online or Portal for ArcGIS, or ArcGIS Pro with the oriented imagery add-in.

Figure 4.3. Perform visual inspections using oriented imagery with the Oriented Imagery widget *(top)* and in Map Viewer *(bottom)*. Both interfaces enable users to select a location on the map to view the best image available, with the image's footprint dynamically updating on the map as users pan and zoom within the image.

Integration with drones

When collecting data, QuickCapture can use the locations from the mobile device, from photos and videos on your phone, or locations from drones. ArcGIS Site Scan, a drone plan and control app, can share a drone's location with QuickCapture. This enables the capture of point, line, and polygon features directly within QuickCapture using the drone's location midflight rather than the location of the mobile device running QuickCapture.

The workflow involves two operators using two tablets: one running Site Scan and the other running QuickCapture. Site Scan is used to guide the drone (and provide video output) while QuickCapture is used to capture records and upload them to ArcGIS. In this integration, the Site Scan app provides a position source to QuickCapture through Bluetooth, allowing QuickCapture to use the drone's location as a location source.

The benefits of such a workflow (figure 4.4) are numerous. First, drones can access locations that are too high, too far, too hot, or too dangerous for people to reach, greatly extending the range of field data collection. Second, information from the drone can be pushed directly into ArcGIS without waiting for the drone to land. Thus, web maps and dashboards can be updated with near-real-time spatial information.

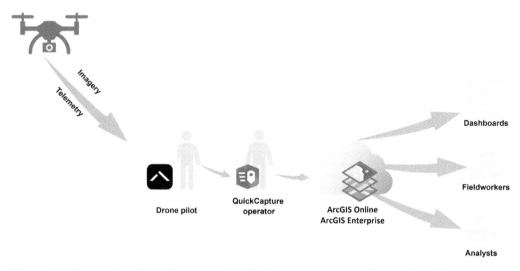

Figure 4.4. Integrating QuickCapture with Site Scan and drones effectively adds "eyes in the sky," significantly extending the reach of data collection.

Tutorial 4: Design QuickCapture projects for rapid data collection and oriented imagery-based inspection

This tutorial comprises two use cases: Use case 1 includes sections 4.1 to 4.4, and use case 2 includes section 4.5.

Use case 1

An organization aims to enable its users, while driving, to report traffic conditions, accidents, police presence, and requests for roadside assistance.

Requirements: The user interface should be easy to use and require minimal user interaction. The solution needs to collect the following two layers:
- A line layer of user's tracks with start time, end time, and geodetic length.
- A point layer to store the following four report types, in addition to the travel speed, direction, and the city of each report:
    - Traffic conditions, such as standstill, heavy, moderate, and other.
    - Crash severity, such as minor or major, and other.
    - Police presence.
    - Roadside assistance requests, such as for battery, tire, medical, gas, and other.

System requirements:
- A mobile device for installing QuickCapture for use in section 4.4.

4.1: Configure the feature layers, styles, and web map

In this section, you will first create two layers by doing the following:
- **Copying an existing feature layer:** This will create an empty layer with the same schema as the source layer. The new layer is separated from the source.
- **Referencing an existing feature layer:** This will register a layer pointing to the source layer, and all data will be stored in the source layer.

By creating layers this way, you can avoid the need to define the attribute fields manually, thereby expediting the tutorial.

After the layers are created, you will then create a web map and style the layers, making the styles ready to import to the QuickCapture project in the next section.

1. On your computer, start a web browser, go to ArcGIS Online (www.arcgis.com or the URL of your ArcGIS Online organization), and sign in.

2. Click Content.

   First, you will create the track line layer by copying an existing layer.

3. Click New item and click Feature layer.

4. In the Create a Feature Layer window, apply the following:
   a. Click Select an Existing Feature Layer and then click Next.
   b. Click My Content and choose ArcGIS Online from the list.
   c. In ArcGIS Online, search for Sample Track lines owner:GTKMobileGIS.
   d. Select the Track Lines feature layer and then click Next.
   e. Click Next. Leave the options as is and then click Next.
   f. In the New Item window, type the title Track Lines (your name). Click Save.

Adding your name makes the item name unique in your ArcGIS Online organization.

After the layer is created, the layer's item page will display with the layer's details.

5. Click Share and share the layer with everyone.

6. Click the Data tab.

On the Table tab, the attribute fields appear. The layer doesn't have any records.

7. Click the Settings tab.

The layer is approved to be shared with the public when editing is enabled.

Sharing the layer with the public can make it easier for you to test the QuickCapture project in the following sections. You can also keep the layer private or share it with only certain groups.

8. Under Editing, do the following:
    a. Check Enable Editing, Keep Track of Who Edited the Data, and Enable Sync.
    b. Uncheck Delete.
    c. Check Editors Can See All Features and Editors Can Only Edit their Own Features.
    d. Click Save.

Next, you will create a layer reference pointing to an existing layer.

98　*Getting to Know Mobile GIS*

9. On the upper menu bar, click the Search button, search for sample road reports layer owner:GTKMobileGIS. Under Filters, click the toggle key to turn off Only Search in <your ArcGIS Online organization name>.

10. In the search result, find the Road Reports layer, click the Options button (three dots), and choose View Details.

    The item page appears, displaying item details.

11. Click the Visualization tab, click the Save As button, set the title to Road Reports (your name), and then click Save.

    This process has registered a layer pointing to its source layer. The item page displays.

12. Click Share and share the layer with everyone.

    Next, you will add the two layers to a web map to configure their styles.

13. On the upper menu bar, click Map to open Map Viewer.

14. On the left toolbar, click the Layers tab.

15. Click Add Layers. Add the Road Reports and Track Lines layers to the map.

    Next, you will add a US cities polygon layer for use in the Arcade script of section 4.3 to query for the cities at the report points.

16. Click the My Content drop-down arrow and choose ArcGIS Online.

17. Search for USA Census Cities owner:GTKMobileGIS.

18. Add the found layer to your map.

19. Click the back button next to Add Layer.

20. In the layers list, drag the USA Cities layer to the bottom of the layers.

21. Hover over the USA Cities layer. Click the Visibility button to turn it off.

    Turning the layer off will improve the map drawing speed in QuickCapture.

    Next, you will style the Road Reports layer by the Report Type field.

22. In the layer list, click the Road Reports layer to make it the current layer.

23. On the right toolbar, click the Styles button and apply the following settings:
    a. For step 1, choose Attributes, click the Add Field button, choose Report Type, and click Add.
    b. For step 2, pick a style and click Types (unique symbols).
    c. Click the symbol for Police to change the style.
    d. In the Symbol Style pane, click the current symbol. In the Change Symbol pane, click the Category arrow and choose an appropriate symbol. Click Done. Set the size to 16 px. Turn off Adjust Size Automatically.
    e. Choose appropriate symbols for Traffic, Crash, and Roadside help.

    > **Note:** You can change the symbol from grid view to list view and then search for the symbols. For example, you can search all the symbols illustrated in the Government category:
    > - Police: Search Security.
    > - Traffic: Search Vehicle.
    > - Crash: Search Traffic.
    > - Roadside help: Search Volunteer.

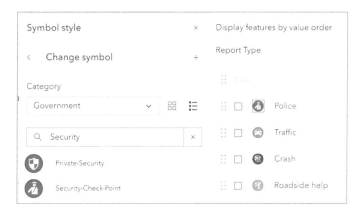

Next, you will save the style to the Road Reports layer itself.

24. With Road Reports as the current layer, on the right toolbar, click Properties.

25. Expand the Information group and click Save.

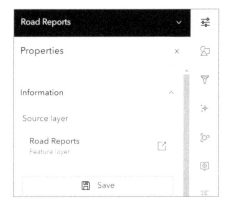

The Track Lines layer uses a default single symbol style. You will save the default symbol to the layer.

26. In the Layers list on the left, click the Track Lines layer to make it the current layer.

27. On the right toolbar, click Properties.

28. Expand the Information group and click Save.

Next, save and share your web map.

29. On the left toolbar, click Save and Open.

30. Click Save As to save the web map. Set the title to Road Reports Map and click Save.

31. Click Share Map, share the map with everyone, and click Save.

    You have created two layers and styled them in the web map. You will use them in your QuickCapture project in the next section.

4.2: Design a basic QuickCapture project
In this section, you will design a QuickCapture project and configure the look and feel of the buttons and the button user inputs.

1. In the upper right, click the app launcher and choose QuickCapture.

    If you closed the window from the previous section, open a web browser, go to ArcGIS Online, sign in, click the app launcher, and choose QuickCapture.

2. Click Add New Project and choose Start from Existing Layers.

3. Under the My Layers tab, click the Road Reports and Track Lines layers you just created to select them.

4. Keep the defaults and click Next.

    With the defaults, the designer will create buttons from symbols of your layers and will not create a new web map.

5. For the project title, type Quick Road Reports. For Save in Folder, create a new folder called Chapter 4. Click Create.

    After your QuickCapture project is created, QuickCapture Web Designer displays. Notice the project preview on the left side and the configuration options on the right side.

    QuickCapture Web Designer automatically creates a group of buttons for each layer in your project. The button symbols for the point layer and the button color for the line layer come from the styles you designed in the previous section.

Next, you will set the shape and camera settings for all four buttons in the Road Reports point layer together.

6. Press and hold the Shift key. In the preview, click each of the buttons in the Point group to select all of them.

7. In the Options panel, under the Appearance tab, apply the following settings:
   a. For Shape, choose the round corner button.
   b. For Colors, choose Fill. Click the color button to choose a light gray or any other color you prefer.

8. While the four buttons are selected, click the Data tab, enable the Show Camera option, and check Hide Camera Preview.

Next, you will configure the data fields for the Crash button.

9. In the project preview, click the Crash button to select this button only. Click the Data tab.
   a. For Capture Fields, Report Type is already set as Crash.
   b. For the Severity field, click the left drop-down arrow and select Button User Input. Click the Select User Input arrow and choose Create New.
   c. In the New User Input window, set the label to Select Crash Severity. Click Create.

Next, you will configure the Roadside help button.

10. In the project preview, click the Roadside help button. Click the Data tab.
    a. Report Type is already set as Roadside help.
    b. For the Help Type field, click the left drop-down arrow and choose Button User Input. Click the Select User Input arrow and choose Create New.
    c. In the New User Input window, set the label to Select Help Type. For Options, select This Is a Required User Input. Click Create.

Next, you will configure the line layer.

11. In the project preview, click the button in the Track group and apply the following settings for Appearance:
    a. Set the label text to New Track.
    b. Set its shape as round cornered.
    c. For Image, click the Add arrow and choose Browse Gallery. Choose an appropriate symbol. For instance, click the Transportation group and choose a symbol for a road line.

Next, you will change the project map to the map you created in the previous section.

12. On the upper menu bar, click the Configure Project Map button and apply the following settings:
    a. Click the Change button.
    b. In the Select a Map window, click My Maps and select the Road Reports Map you created in the previous section.
    c. For Display on Tablet, choose Show Buttons and Map Side by Side.

13. On the ribbon, click Save to save your project.

You have created a basic QuickCapture project. You can continue to the next section to enhance it or skip to section 4.4 to publish it and start collecting data.

4.3: Add device variables and Arcade calculations

In this section, you will use device variables and Arcade to populate additional fields automatically without users having to manually input them.

You will first configure the start time, end time, and geodetic lengths of the line features.

1. In the project preview, click the New Track button for the Lines layer.

   If you have closed the project, go to ArcGIS Online, sign in, click the app launcher, click QuickCapture, and then click the project to open it in Web Designer.

2. In the Options panel, click the Data tab and set the following:
    a. For the Start Time field, click the drop-down arrow and choose Device Variable. Click the Select Variable arrow and choose Start Time.
    b. Set the End Time field to use the End Time device variable.

3. For the Length field, click the drop-down arrow and choose Arcade Expression. Click the Select Expression arrow and click Create New.

4. In the New Arcade Expression window, for Label, type Calc Geodetic Length.

5. Click the *fx* button on the window's right toolbar. Search for length. In the search result, under Geometry Functions, notice LengthGeodetic. Click the Expand arrow to learn more about it.

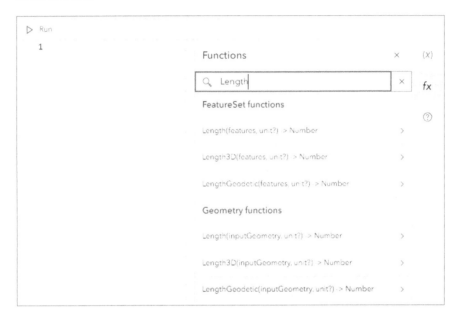

6. Read the description of the function and consider the two parameters it needs. Scroll down to see the example.

7. Select the example `LengthGeodetic($feature, 'kilometers')` and copy and paste it to the script area.

   You can also copy the code from https://arcg.is/99qf.

8. Click the Run button to test it.

   The output is 0 because there are no lines in the track layers yet.

9. Click the Create button.

   Next, you will configure the four buttons of the point layer.

10. Press and hold the Shift key. In the preview, click each of the buttons in the Point group to select all of them.

11. In the right panel, click the Data tab and set the following:
    a. For the Speed field, click the drop-down arrow and choose Device Variable. Click the Select Variable arrow, click Travel, and click Speed (km/h).
    b. For the Direction of Travel field, click the drop-down arrow and choose Device Variable. Click the Select Variable arrow, click Travel, and click Direction of Travel.
    c. For the City field, click the drop-down arrow and choose Arcade Expression. Click the Select Expression arrow and click Create New.

12. In the New Arcade Expression window, for Label, type Calc City.

13. In the script text area, copy the following code from https://arcg.is/99qf:

```
var cities = FeatureSetByName($map, "USA Cities", ["NAME"]);
var city = First(Intersects($feature, cities));
if (!IsEmpty(city)) {
 return city["NAME"];
} else {
 return null;
}
```

The script first gets the Cities layer from your web map and then uses the report point to intersect with the Cities polygon layer. If a city is found, it returns the city's Name attribute value.

14. Click Create.

Next, you will write an Arcade script to calculate the traffic condition if the Traffic button is tapped.

15. In the project preview, click the Traffic button to select it.

16. In the right panel, click the Data tab. For the Traffic Level field, click the drop-down arrow and choose Arcade Expression. Click the Select Expression arrow and click Create New.

17. In the New Arcade Expression window, for Label, type Calc Traffic Condition.

18. In the script text area, copy the following code from https://arcg.is/99qf:

    ```
    if (IsEmpty($feature.Speed) || $feature.Speed < 5) {
     return "Standstill"
    } else if ($feature.Speed < 15) {
     return "Heavy"
    } else if ($feature.Speed < 35) {
     return "Moderate"
    } else {
     return "Other side"
    }
    ```

    The script evaluates the speed attribute of the report feature and returns traffic condition classes based on the speeds.

19. Click Create.

20. Click Save to save your project.

    You have configured your project to automatically collect additional fields without requiring your users to type them in.

4.4: Collect data using the QuickCapture field app

In this section, you will review and preview your project in QuickCapture Web Designer and publish it. You will then open the project in the QuickCapture field app to collect some data. Finally, you will review the data collected on your web map.

> Note: If you didn't create the project, you can skip to the step for installing QuickCapture and use the provided QuickCapture project to complete the rest of the steps in this section.

1. On the upper menu bar, click the General Settings button.

    If you have closed the project, go to ArcGIS Online, sign in, click the app launcher, click QuickCapture, and then click the project to open it in Web Designer.

2. Read about the following settings.
   - **Required Accuracy:** If this accuracy threshold is not met, the app will prevent data capture. For indoor environments, your phone's location accuracy is generally low. It's advisable to leave the location setting unset so that you can successfully test data capture while indoors.
   - **Distance Threshold:** This defines the minimum distance that must exist between two consecutive vertices or points. This applies to capturing streaming points, lines, and polygons.
   - **Allow location edits and time limit:** This setting controls whether you allow users to edit the locations and set a time limit after capture.

3. On the upper menu bar, click the Change Device Preview button to preview the project.

4. Switch between phones and tablets, in both portrait and landscape modes.

   In tablet landscape mode, the buttons and the map display side by side in the split-screen view.

5. Click Share. Share your project with everyone or certain groups.

   The option to share with everyone is available only to organizations with ArcGIS Hub Premium.

6. Click Share again. Read the Sharing options, including URL Link and QR Code. Click the QR Code button to display your project QR code.

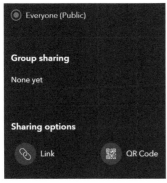

You can right-click the QR code, save the image, email the image to your users, or embed it in your organization's website. Your users can scan the QR code without having to type a long URL.

Next, you will use the field app to capture data.

7. On your mobile device, install QuickCapture if you don't yet have it.

   QuickCapture is available from the Apple App Store, Google Play, and the Microsoft Store.

8. Scan your QuickCapture project QR code or open the ArcGIS QuickCapture app and sign in if your project is not shared with the public.

   > **Note:** If you didn't create the project or if you prefer to use the project provided, you can scan the QR code illustrated above.

9. If this is the first time you have opened QuickCapture, allow access to your location, camera, and Bluetooth if prompted.

10. If you scanned the provided QR code, your project should open automatically. Otherwise, choose Browse Projects, find the Quick Road Reports project you created earlier, and tap it.

    After the project is downloaded and is opened, you will begin to collect a line.

11. Tap New Track to start collecting a line feature. If possible, walk out of your office or classroom to do so, because it requires a minimum of 10 meters to collect a line.

    You hear "Start New Track," spoken by the app. The app starts to collect the line feature using streaming.

    While the track line is being collected, you will collect several point features.

12. Aim your device at an object and tap a report button (for example, Crash) and answer the relevant question that appears.

Chapter 4: Rapid data collection using ArcGIS QuickCapture

You should hear a camera click and the report type spoken. The point is captured at your current location, with the report type recorded and a photo attached. All these actions are completed with one button push.

13. Notice the Edit and Delete buttons that appear. You can edit or delete the data by tapping them within six seconds of data capture.

    The data captured is stored temporarily on your device for 30 seconds by default until it is sent to the feature layer in ArcGIS Online. This default duration can be changed in the QuickCapture mobile app (Settings > Autosend).

14. Within 30 seconds of capturing the data, tap the Map button in the upper-right corner.

    The map will display.

15. Tap a report point you just captured to review the location, the photo, and delete the record if needed.

16. Repeat the preceding steps to collect one point for each of the remaining report types.

17. Walk for a distance and tap New Track again to stop streaming.

    You will hear "End New Track" spoken by the app.

    Next, you can review the data you collected.

18. In a web browser, sign in to ArcGIS Online and open the QuickCapture project web map you created in section 4.1.

    > **Note:** If you used the provided QuickCapture project to collect data, you can review the data collected using the web map at https://arcg.is/1WyvTv0. You will be able to see only your own data. You may delete your own data if you like.

19. Click the points and lines; observe the attributes from user inputs, device variables, and Arcade calculations; and review the photos collected for the point reports.

Use case 2

An organization seeks to improve its asset inspection workflow using oriented imagery.

Requirements: The solution should include a mobile app and a web app.
- **Mobile app:** The app for mobile workers should facilitate easy photo capture of assets and automatically calculate image metadata.
- **In-office web map or app:** This tool should enable reviewers to examine the photos and their geographic coverages on a map, supporting thorough asset inspections and helping to identify the locations and conditions of assets requiring maintenance.

System requirements:
- An ArcGIS Online creator or publisher account.
- A mobile device for installing QuickCapture for use in section 4.5.

4.5: Inspect assets by integrating QuickCapture with oriented images

This section will guide you through quickly creating a QuickCapture project for collecting oriented images using a template. Here is an introduction to this template: https://arcg.is/1K4ir11. For instructions on creating a project from scratch without using the template, refer to this tutorial: https://arcg.is/LviDH.

1. Open a web browser, navigate to ArcGIS Online (www.arcgis.com or the URL of your ArcGIS Online organization), and sign in.

2. Click the app launcher and choose QuickCapture.

3. Click Add New Project and choose Start from Template.

4. Search for Photo Inspection.

5. In the search results, click the Photo Inspection template.

A preview and introduction page will appear.

6. Read the introduction and click the Use Template button.

7. In the Deploy Template window, add your name to the title and click Create.

    The deployment will create all the ArcGIS items included with the template and adds them to a new folder in your content.

    After the project is deployed successfully, you will notice two buttons, Edit Project and View Template.

8. Click View Template.

    The item page of the solution appears in ArcGIS Online.

9. Expand the Solution Contents tree. Review the content items created by the template and consider the relationship between the items.

At the root level, the content item hierarchy tree shows that the template created a QuickCapture project named Photo Inspection and a Photo Inspection Dashboard for desktop view.

The QuickCapture project includes the following items:
- A URL link to a Photo Inspection Dashboard designed for viewing on mobile screen sizes. The dashboard uses the dashboard view of the Photo feature layer. This view is read only.
- A mobile view of the Photo feature layer. This view is enabled for editing. For security purposes, editors can add features only through this view. they can't delete or see any features, even those they added.
- A project map is named Photo Inspection Map, which includes the dashboard view of the Photo layer.
- An OI catalog named Photo_OIC, which is based on the Oriented Image Catalog (OIC) view of the photo layer.

10. Click the app launcher and choose QuickCapture.

11. Click the Photo Inspection project that you just created.

The project opens in Web Designer.

12. Consider the following points about the design of the project.
    - It has a project user input asking for the inspector's name. Users must answer this question when they first open the project.
    - It has a group with four buttons for users to specify the severity of the problem and to take photos. Note the project uses photo locations instead of phone locations for captured records.
    - It has a link button, linking to the mobile view of the Photo Inspection Dashboard. This dashboard allows users to view the photos.

13. On the upper menu bar, click the Manage Project Layers button.

14. Click the Options button for the Photo layer. Click View Oriented Imagery.

    Your OIC will open in an OIC viewer. You will be prompted to sign in because your OIC is not shared with the public.

15. Sign in.

    Once you have collected some photos using QuickCapture, you can access this URL to view the OIC.

16. Go back to your QuickCapture project and click Share to share the project.

17. Expand the warning messages on the left toolbar. Click the link to share the mobile view of the dashboard with everyone.

    Next, you will collect data in the field, using your own project or using the provided project if you prefer.

18. On your mobile device, scan the QR code here to use the QuickCapture project provided or scan the QR code of your own project.

The Photo Inspection project downloads and opens.

> **Note:** If you use the provided project, ensure that the photos you take do not compromise your privacy. For enhanced privacy and better location accuracy, it is advisable to take photos outdoors. Photos collected in this project will be regularly deleted to maintain privacy.

19. Collect five or more photos for one or two assets.

    The quality of the OI data you collect depends on various factors.
    - **Location accuracy:** Accurate location data is essential for precise photo footprint calculations. For better results, it is recommended to collect data outdoors unless you have an indoor positioning system. Using an external GNSS receiver can significantly improve location accuracy.
    - **Photo angles:** Professional use of oriented imagery requires accurate readings of the camera's heading (True North Azimuth), pitch, and roll angles. The best-oriented imagery photos are taken of objects directly in front of you at roughly a 90-degree angle. Avoid taking photos looking down at your feet or up at the sky.
    - **Phone orientation:** Keep your device in either portrait or landscape mode, avoiding any angles in between. This ensures more accurate photo coverage calculations.

    Next, you will perform photo-based inspection using the OIC viewer on your desktop computer.

20. If you used the project provided, on your desktop computer, navigate to https://arcg.is/1fyKuT1. If you used your own project, open the Oriented Imagery viewer of your project (refer to steps 14 and 15).

21. On the map, pan to your area and observe the photo coverages. Click a location on the map to view the photos in the OIC widget.

    A red X marks the clicked point on the map and in the photo.

22. Zoom in, drag the photo, and zoom out to see the current coverage on the map change in sync.

23. On the toolbar of the OIC widget, click the Navigation tool.

    The tool has a compass depicting the camera locations for all the images relevant to the location you clicked.

24. Click a blue dot in the Navigation tool to view a different photo.

25. On the toolbar of the OIC widget, expand the options and click the Image Enhancement tool to adjust the photo brightness, contrast, and sharpening for better inspection of the photos.

In this tutorial, you experienced the workflow of QuickCapture for rapid data collection. In the first use case, you initially created two layers by copying and referencing existing layers, styled them, saved their styles with the new layers, and created a web map. You then created a QuickCapture project from the two layers, which brought their symbols into the QuickCapture buttons. You used the web map as the project's web map. After configuring the appearance of the buttons, you configured the photos, button inputs, device variables, and Arcade scripts. After publishing the project, you used the QuickCapture field app to collect data. As you experienced, the large buttons made it easy to click. The configuration and Arcade scripts made the process rapid and semiautomatic, without the need for manually pinpointing locations on the map or filling out lengthy forms.

In the second use case, you created a QuickCapture project using a template. Instantly, the process created the point layer, layer views, OIC, web map, QuickCapture project, and dashboards for mobile and desktop screen sizes automatically. Once again, field data collection was quick and effortless. The image metadata was calculated as you collected photos, seamlessly. The collected photos support close-in asset inspection across multiple photos and on maps.

Assignment 4: Create a QuickCapture project for rapid data collection

Requirements:
- The project must include a point layer and a line layer.
- Each layer should have a field for resource or asset type with coded values, such as tree, table, playground, trash can, and light poles for the point layer and trail, walkway, and fence for the line layer.
- The layers need to have consistent symbols in a web map and in the buttons of the QuickCapture project.
- The project must collect photos, capture at least two device variables, and include one Arcade script.
- Your project and the content items involved must be shared with the public or a group to which your instructor belongs.

What to submit:
- The URL or QR code to your QuickCapture project
- The URL to your project web map
- Your Arcade script

Chapter 5
Mobile workflow in offline mode

Objectives
- Grasp the necessity of offline support.
- Explore offline support options provided by Field Maps.
- Validate layers for offline compatibility.
- Create preplanned and on-demand offline areas.
- Copy (sideload) basemap packages to your device.
- Use sideloaded basemaps in offline mode.
- Enable the inbox using Survey123 Connect.
- Use the outbox and inbox functionalities in Survey123 mobile app.

Introduction

Previous chapters explored data collection and field awareness in online mode. However, network availability cannot be guaranteed everywhere at all times. Offline support is crucial for enhancing user experience and is particularly vital for mission-critical applications. To ensure Mobile GIS functionality when disconnected, it is necessary for Mobile GIS mobile workers to download data and maps while connected and to synchronize the data upon reconnection. This chapter introduces the options for offline support provided by Field Maps and explains the concepts of outbox and inbox in Survey123. The tutorial sections demonstrate how to validate layers for offline use, create preplanned and on-demand offline areas in Field Maps, enable the Inbox feature using Survey123 Connect, and manage the outbox and inbox in the Survey123 mobile app. Additionally, the tutorials cover how to copy map packages to your device, as well as how to use sideloaded basemaps in offline mode in both Field Maps and Survey123.

The need for offline workflows

Before satellite-based communication becomes affordable, communication between mobile clients and servers primarily depends on Wi-Fi and mobile networks. Wi-Fi, though prevalent in homes, offices, and various public hot spots, is not universally accessible. Mobile networks are primarily built for densely populated regions but often fail to provide reliable coverage in remote areas, which may lack connectivity altogether or have only slow network services. Additionally, even in regions with robust data services, network reliability is not guaranteed 100 percent all the time because there can be outages and maintenance activities.

Mobile GIS often needs to operate in areas with limited or no network connectivity. To perform effectively, it requires access to GIS data, including asset locations, details, and histories while offline. Therefore, the capability for Mobile GIS to operate offline is not merely an advantage but a necessity in many cases. For a Mobile GIS app to operate offline, it should be able to do the following:
- Download the data when there is network access.
- Remain usable without a reliable network connection.
- Present users with local data instead of waiting for the network call to complete or fail.
- Sync the data by uploading a user's local changes and downloading the changes by other users when a data connection is available.

ArcGIS Mobile GIS apps, including Field Maps, Survey123, and QuickCapture, are equipped with robust offline capabilities, ensuring that field operations remain uninterrupted, regardless of network conditions.

Offline map areas and mobile map packages for Field Maps

With Field Maps, you can package the operational layers, basemap layers, and attachments of a map for offline use by creating offline map areas or by creating mobile map packages (MMPKs). MMPKs, authored in ArcGIS Pro, are for downloading and viewing data only, not for data collection. This chapter will focus on offline map areas.

Offline map areas enable you to package your data, basemaps, and attachments for download. Once these areas are downloaded to a mobile device, users can view existing data and collect new data offline, similar to the connected experience. Once connected again, mobile workers can synchronize their data with the server or cloud. The two main methods for creating map areas are as follows.
- **Preplanned areas using Field Maps Designer:** Map authors can prepare maps for areas known to have connectivity issues in advance of where mobile workers will be operating. Once created, these map areas are packaged and made available for download in the Field Maps mobile app to anyone with whom the map is shared. This preferred method allows map areas to be prepared and packaged once, then downloaded by multiple mobile workers. It is a "one for all" method, saving mobile workers from having to define the areas themselves.
- **On-demand offline areas using the Field Maps mobile app:** For unplanned or specific offline tasks, mobile workers can independently create and download a map area directly on their device using the Field Maps mobile app. these map areas are intended for single use. This "one for one" method means that if another mobile worker needs the same map area, they must define the area and create the online package themselves.

Before you can create offline map areas, your web map needs to be enabled for offline use, which requires that all layers support offline uses. Field Maps Designer conducts a validation check on all layers in the web map. Common issues identified during this validation process are illustrated in figure 5.1.

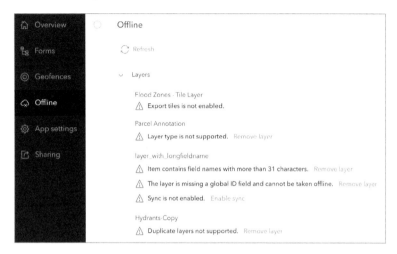

Figure 5.1. Field Maps Designer automatically performs validation checks and provides meaningful error messages to help map authors ensure all layers in the web map support offline use.

When creating offline areas, Field Maps Designer allows you to choose whether to include attachments in the packages. Attachments, typically larger than vector geometry data, can substantially increase data transfer volumes. By default, all features and attachments for both editable and read-only layers are packaged for downloading and will be synced. However, not all this information may be necessary for completing the field tasks. To enhance performance and reduce data transfer costs—especially in areas with low bandwidth—map authors may choose to exclude attachments. By adjusting the offline options, map authors can control the types of data retrieved. Crucially, these settings do not affect the edits made by mobile workers, which are always sent to the server.

Copy (sideload) and reference offline basemaps

When configuring web map offline settings in Field Maps Designer, you can choose to use a sideloaded basemap. The main reason for using a sideloaded basemap instead of the basemap in the package coming with the offline map area is often about the size and complexity of the basemaps. When creating offline areas, you're constrained by basemap size limits (figure 5.2), which cap at 150,000 raster tiles and 10,000 vector tiles unless you supply your own basemaps that support export tiles. For raster tiles, each deeper level of detail (LOD) quadruples the number of tiles compared with the previous level. Vector tiles, however, are divided based on point/vertex density; although the increase in the number of tiles isn't as steep, it's still significant. Therefore, when selecting a large area, you may need to compromise on the LOD to remain within these limits. Conversely, if the deepest LOD is necessary, you'll need to reduce the geographic size of the offline area to ensure the total package size stays within allowable constraints.

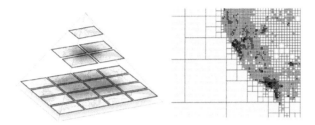

Figure 5.2. For raster tiles *(left)*, each deeper detail level quadruples the number of tiles from the previous level. For vector tiles *(right)*, which are divided by point density, the increase rate is less steep but can still be significant.

Basemaps copied to your device offer a way to bypass the usual size limits imposed on offline areas. these basemaps include file types such as TPKX (tile package) for raster tiles and VTPK (vector tile package) for vector tiles. Sideloading involves transferring these files to specific directories on mobile devices: for Field Maps, this is Field Maps/Basemaps on iOS and \Android\data\com.esri.fieldmaps\files\basemaps on Android.

Once the basemap files are placed in these directories, they appear in the basemap gallery. Mobile workers can then select these basemaps from the gallery, offering an experience like that of a connected environment. It's important to ensure that the spatial reference of the default web map basemap matches that of the local basemap. Additionally, the offline map's extent must intersect with that of the local basemap for proper functionality.

In addition to sideloading basemaps, Field Maps Designer allows you to reference a basemap from your organization to simplify deployment and sharing. With this approach, you can share the basemap with every mobile worker who uses the map. Once downloaded, the basemap can be reused across multiple web maps, eliminating the need for repeated downloads.

Survey123 can also use sideloaded basemaps, including TPKX, VTPK, and MMPK. You can copy basemaps to a designated folder:
- Survey123/Documents/ArcGIS/My Surveys/Maps on iOS, and
- Android/data/com.esri.survey123/files/ArcGIS/My Surveys/Maps on Android.

Additionally, using linked content allows you to link a basemap file, which automatically places the file in the correct folder upon download, streamlining the process.

Survey123 inbox and outbox

Survey123 supports offline operations through inbox and outbox functionalities (figure 5.3). Note that Inbox and Outbox are available only in the Survey123 field app, not in the browser app.

- **Inbox:** The Survey123 inbox allows you to download existing data from the survey layer while connected. Once the data is downloaded, you can review and edit the locations and details of existing features even when disconnected.
- **Outbox:** When working disconnected, the Survey123 field app automatically saves new surveys into an outbox, similar to an email outbox. these can be submitted once a network connection is available.

Figure 5.3. Survey123 supports offline operations through its inbox and outbox functionalities. When connected, you can download data to the inbox (*middle*) and send data stored in the outbox (*right*). When disconnected, you can continue to edit existing data and collect new data (*left*).

The two important considerations when using Survey123 for offline operations are as follows.
- **Basemaps for your survey layer:** To allow users to view points, lines, and polygons on a basemap or to draw geometries on basemaps, basemaps must be accessible offline. You can ensure this by copying the basemaps to your device or using linked content to provide them.

- **Offline functionality limitations:** Questions that require live access to web services, including feature layers, will not function offline. This includes features such as pulling data from layers, search appearance, geocode appearance, and JavaScript functions that rely on feature services or other web services, which will all be inoperative in offline settings.

Tutorial 5: Design offline workflows with Field Maps and Survey123

In this tutorial, you will enhance the hydrant inspection solution built in Field Maps in chapter 2 and the issue reporter solution built in Survey123 in chapter 3 and make them work in offline mode.

Functional requirements:
- Facilitate situational awareness and inspection using Field Maps in offline mode. Allow field workers to
    - review the existing hydrants and their inspection history,
    - perform new inspections for existing hydrants,
    - collect new hydrants,
    - have basemap to use even when not connected, and
    - sync data by uploading and downloading when connected.
- Facilitate situational awareness and resolving the issues workflow using Survey123 in offline mode.
- Allow the public to report new issues in offline mode.
- Allow facilities field workers to
    - review the issues received in list and on map,
    - search for existing issues and review issue details,
    - update existing issues and status,
    - access basemap even when not connected, and
    - sync data up and down when connected.

System requirements:
- Section 5.4 requires a Windows computer to run Survey123 Connect. other sections require a browser and a mobile device.

Chapter 5: Mobile workflow in offline mode   **123**

5.1: Validate layers for offline use in Field Maps Designer
In this section, you will validate your layers for offline use and enable offline use of your map.

1. In a web browser, navigate to ArcGIS Online and sign in.

2. On the top toolbar, click the Search button and search for Offline Validation sample map owner:GTKMobileGIS.

3. Under Filters, turn off Only Search in <your ArcGIS Online organization name>.

4. Click Open in Map Viewer.

   This is a modified version of the Hydrant Inspection web map you created in chapter 2, with two additional sample layers.

5. Expand the Contents toolbar and click Save and Open > Save As. Save the map as Hydrant Inspection Offline in the folder Chapter5.

6. In the upper right of the screen, click the app launcher and click Field Maps Designer.

7. Click the Hydrant Inspection Offline web map to open it.

8. On the Contents toolbar, click Offline.

   At the top of the Offline page, the Offline toggle key of the map is made unavailable and cannot be enabled.

   Your web map is automatically checked for offline compatibility. It currently displays several issues, as outlined in the error and warning messages. these include field names exceeding 31 characters, absent global ID fields, sync unavailable, and duplicated layers. For the map to be suitable for offline use, all layers, tables, and basemaps listed in the Content section must be enabled for offline access. All errors need to be addressed, whereas resolving warning messages is recommended but optional.

9. Read the error message for the layer_with_longfieldname layer:
   - Item contains field names with more than 31 characters.
   - The layer is missing a global ID field and cannot be taken offline.
   - Sync is not enabled.

   If you owned the layer, you would have the option to delete or rename the long field name and enable synchronization, which would automatically generate a global ID field. However, to simplify this tutorial, you should remove the layer.

10. For layer_with_longfieldname, click Remove Layer.

    The Hydrants-Copy and Hydrants layers both have the Duplicate Layers Are Not Supported error.

11. For Hydrants-Copy, click Remove Layer.

    The Offline toggle key of the map is now enabled. With offline enabled, your web map will work for both online and offline uses.

    By default, all features and attachments for both editable and read-only layers are downloaded. To speed up the tutorial, you will exclude attachments from the downloads.

12. On the Offline page, click the Features and Attachment section to expand it.

13. For Editable Features, click Features Only.

14. Click Save.

5.2: Create preplanned offline areas

Next, you will first explore the limits of the combination of area size and level of detail and then create an offline area.

1. On the Offline page, click the Map Areas section to expand it and click Manage Areas.

2. Click Create Offline Area.

   The sketch tools and a map display.

3. Click the Sketch Rectangular Map Area button (rectangle with a plus sign).

Your cursor changes to indicate that it is now in sketch mode.

4. On the map, draw a rectangle large enough to encompass the size of several states.

   Depending on the size of the area you drew, you may see a warning message saying, "Offline area is too large. Reduce the area or the level of detail."

5. Click the Level of Detail section to expand the options:
   - If you see the too-large warning, drag the right slider knob to the left to reduce the LOD and watch the warning message disappear.
   - If you do not see the warning, drag the right slider knob to the right to increase the LOD and observe the warning message appear.

   For raster tile basemaps, each increase in the LOD results in quadrupling the number of tiles compared with the previous level. To streamline and expedite this tutorial, you will create a small offline area encompassing the southern part of Citrus Plaza in Redlands, California.

6. On the map, click the offline area you drew to select it. Press the Delete key to remove it.

7. Click Sketch Rectangle Map Area to exit the draw mode.

8. In the map search bar, type Citrus Plaza South, Redlands, CA.

   The map zooms to the area.

9. Close the search result pop-up.

10. Click Sketch Polygon and draw a polygon that encompasses the plaza. For west, south, and east boundaries, use Alabama Street, West Lugonia Avenue, and Citrus Plaza Drive, respectively. For the north boundary, make sure you include the buildings in your polygon.

11. Set the Level of Detail so that it ranges from Town to Building level.

12. Name the package Citrus Plaza South.

13. Click Save.

The offline map area will be packaged, which can take several minutes. After the package is created, you may click on it to see the size of each content type included.

14. In the upper right of the Manage Areas window, click Close.

    The Map Areas section displays the number of offline areas in your map.

5.3: Use preplanned areas and create on-demand offline areas in the Field Maps app

1. On your mobile device, start Field Maps and sign in.

2. Locate the Hydrant Inspection Offline web map you created in the previous section and open it.

    The map and its offline areas are listed, including a card for the offline area you created in the previous steps.

    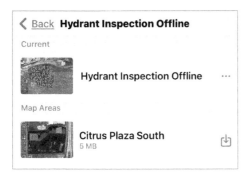

3. Tap the Citrus Plaza South map card to download it.

    Once downloaded, you can interact with the map as you would with an online map, both with and without a data connection.

    Next, you'll switch to offline mode to explore the map and collect data.

4. On your mobile device, enable Airplane mode and turn off Wi-Fi.

5. In the Maps list, tap the Citrus Plaza South offline area to open it.

6. Enable Location Alerts.

7. Tap a hydrant in the area to review its attributes and previous inspections.

8. Edit the hydrant or add an inspection and tap Submit.

   At the top of the screen, a dot appears under the upward arrow of the Sync button (⇅), indicating that you have local data to sync with the cloud.

9. On your mobile device, turn off Airplane mode and turn on Wi-Fi.

   A sync option appears—Sync Now or Auto-sync at a Time Interval.

10. Click Sync Now.

    Next, you will create an on-demand offline area for the north part of Citrus Plaza, Redlands.

11. Tap the back button to return to the list of map areas.

12. Tap the Overflow menu (three dots) next to Online Map and tap Add Offline Area.

    The Add Offline Area screen appears, showing a rectangle outlining the maximum area possible at the default level of detail.

13. Pan and zoom the map to any area you prefer—for example, your current area—and adjust the LOD as needed.

    Selecting a small area will expedite the creation process.

14. Tap Download Area.

    You are returned to the list of maps, which includes a map card indicating that the offline area you just created is being generated and downloaded.

15. Once the download completes, tap it to open it.

    You can use it the same way as the offline areas created in the previous section using Field Maps Designer.

16. Optionally, you can switch your device to offline mode, capture data for a hydrant, and then switch back to online mode to synchronize your data, following the process given in previous steps.

5.4: Copy and use the VTPK basemap in the Field Maps app

In addition to using basemaps from offline areas, you can copy or reference separate vector tile map packages or tile map packages as basemaps. This option allows the use of larger, more detailed basemaps with specialized symbology for offline use.

1. If you use iOS:
   - On your mobile browser, navigate to https://arcg.is/1WraGD (Vector_Basemap_Citrus_Plaza.vtpk, 23M) and tap Download. Ignore the Unsupported File Type warning.
   - After the file is downloaded, tap Open In and choose Save to Files, browse to Field Maps, and tap Save.
   - Continue to the step for starting the Field Maps app.

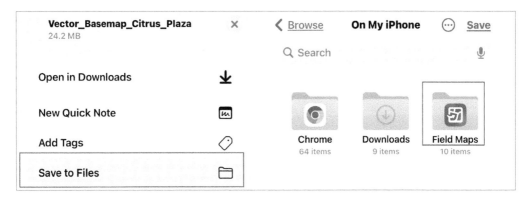

2. If you use Android:
   - On your desktop computer, navigate to https://arcg.is/1WraGD (Vector_Basemap_Citrus_Plaza.vtpk, 23M) and tap Download.
   - After the file is downloaded, plug your device or SD card into your computer. Using a file navigation tool on your computer, browse to \Android\data\com.esri.fieldmaps\files\basemaps (if this folder doesn't exist, create it).
   - Copy the downloaded basemap file to the Basemaps folder.

   Next, you will use the basemap in the Field Maps app.

3. On your mobile device, start the Field Maps app and sign in.

4. Find and open the Hydrant Inspection Offline map.

   You will use the VTPK in offline mode.

5. Enable Airplane mode and turn off Wi-Fi.

6. Open the Citrus Plaza offline area.

7. Tap the Overflow menu in the upper right of the screen. Tap Basemap.

8. In the Basemap gallery, find and select the VTPK you downloaded.

   The basemap switches to the one you selected.

9. Zoom in to explore the level of detail.

10. Optionally, edit a hydrant or submit an inspection.

11. Turn off Airplane mode and turn on Wi-Fi.

5.5: Enable the inbox and link the offline basemap in Survey123 Connect (optional)

This section requires a Windows computer, as required by Survey123 Connect.

1. Start Survey123 Connect and sign in with your ArcGIS Online account.

2. Download the Excel file from https://arcg.is/1Tumb10. Drag it into Survey123 Connect. Click OK after the survey is created.

   This will generate a simplified version of the survey described in section 3.6. The dynamic calculation of campus name and generation of issue list are removed since they require an internet connection.

   Before enabling the inbox, you will examine the Excel workbook to understand the settings for linking to existing data and retrieving data from the related table.

3. On the toolbar, click the XLSForm button to open the Excel workbook.

4. In the Excel workbook, click the Survey tab. Locate the Begin Repeat row and scroll to the right to the bind::esri:parameters column. Notice that the value is query.

   The Query option will return all repeats—in other words, records from the related table—to the inbox. these records are read-only. Users can review existing related records and can add new ones but can't edit existing records. In situations where editing is needed, the value would need to be `Query, allowUpdates=True`.

5. In the Excel workbook, click the Settings tab.

   The submission_url points to an existing feature layer. Existing data in the layer can be downloaded to the inbox. The Settings tab also includes the instance_name setting, which defines the display label for each record in the inbox.

   The instance_name is `concat(${issue_type}, " - ", ${level_of_urgency})`.

   The formula will show both the issue type and level of urgency as the label for each issue in the inbox.

   Next, you will enable the inbox.

6. In Survey123 Connect, on the bottom toolbar, click Options.

7. For Inbox, turn on Enable Inbox.

   A warning appears indicating that when both the inbox and the sent box are enabled, refreshing the inbox will not download responses currently in the sent box, which can be confusing to mobile workers. Therefore, you should turn off the Sent folder.

8. For Sent, turn off Enable Sent Folder.

9. On the left toolbar of Survey123 Connect, click Publish.

   A message will appear to indicate that the survey is configured to use an existing feature layer, rather than creating a new one.

10. Click Publish Survey.

11. After the survey is published, click OK.

Next, you will link an offline basemap to the survey, enabling mobile workers to download the basemap directly to the appropriate folder in the Survey123 app.

12. On the lower toolbar, click Linked Content and then click Link Content.

13. Click Map Package.

14. Under Filters, turn off Only Search for Items Owned by <your ArcGIS Online organization name>. Search for Vector Citrus Plaza owner:GTKMobileGIS.

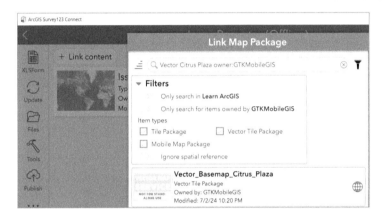

15. Click the VTPK in the result and click OK.

16. Click Publish and click Publish Survey.

17. After the publishing is complete, click OK.

18. Optionally, you may share the survey with everyone.

5.6: Copy the VTPK basemap and use the inbox in Survey123

1. You can use the provided survey or your own survey created in the previous section.
    - **Use the provided survey:** On your mobile device, scan the QR code below or navigate to https://arcg.is/LS5eX0. Sign in to your ArcGIS Online account. The Issue Reporter (Offline) survey downloads and the form opens. Tap the Close Survey X button and then tap Close and Lose Changes.

- **Use your own survey:** Start Survey123 on your mobile device and sign in. In the upper right, click your account button. Click Download Surveys. Locate the survey you created in the previous section and tap Download. Click the back button on the ribbon.

2. From the gallery of downloaded surveys, tap your Issue Reporter (Offline) survey to open it.

   Next, you will download the offline basemap linked to the survey.

3. In the upper right of the screen, tap the menu button (three lines) and tap Offline Maps.

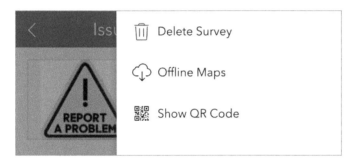

4. Locate Vector_Basemap_Citrus_Plaza VTPK and tap Download.

5. After the download completes, tap the back button.

   This approach uses the linked content approach to copy the basemap to your device. Alternatively, you can copy the VTPK directly to Survey123.

6. Tap Inbox and then tap Refresh.

The inbox fills with a list of existing features in the layer, labeled with the issue type and level of urgency, as configured in the instance_name parameter of the XLSForm. The direction and distance to each issue location are displayed alongside the features.

7. Tap Map.

    The issues are displayed on the map so you can see where they are.

8. In the search bar, search for Citrus Plaza, Redlands and select it from the result list.

    The map zooms to Citrus Plaza.

    Since both the feature layer and the basemap have been downloaded, you can now explore existing data and collect new data in offline mode.

9. Enable Airplane mode and turn off Wi-Fi.

10. Tap Basemap.

11. In the Basemap gallery, tap the Vector_Basemap_Citrus_Plaza VTPK you loaded.

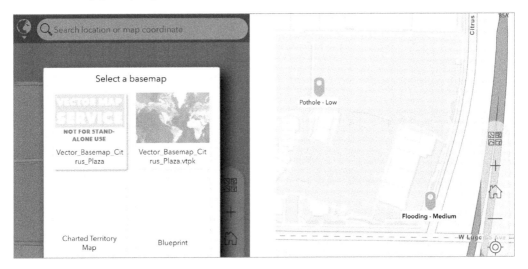

12. Zoom in on the map to see more map details.

13. Tap an issue on the map to open it.

The survey form loads with the issue details. Attached images are not downloaded.

14. Scroll down to the bottom of the form. Click the Add button to add a repeat form and set a status, such as Assigned.

15. Tap Submit and then choose Save in Outbox.

16. Tap the back button.

There is one record in the outbox.

17. Optionally, tap Collect to collect an issue, submit it, and save it in the outbox.

Next, you will reconnect to the internet, submit all surveys, and synchronize your inbox.

18. Turn off Airplane mode and turn on Wi-Fi.

19. Go to the outbox and tap Send in the lower right to submit your surveys.

20. Return to the inbox and tap Refresh.

The new issues you and others submitted have been downloaded.

In this tutorial, you prepared your map and survey for offline use and explored their functionalities using Field Maps and Survey123, respectively, while in Airplane mode with Wi-Fi turned off.

In the Field Maps tutorial sections, you configured the hydrant inspection web map for offline use. Initially, you assumed the role of a content creator, validating layers for offline functionality, creating preplanned areas, and configuring offline settings in Field Maps Designer. Subsequently, you switched to the role of mobile workers. Using the Field Maps mobile app, you downloaded the offline areas, created on-demand offline areas, copied a VTPK basemap to your device, reviewed existing data, and collected and synchronized new data.

In the Survey123 tutorial sections, you set up the Issue Reporter survey for offline operation. As the survey author, you enabled the inbox using Survey123 Connect and linked the survey to a VTPK basemap. Switching to the role of a mobile worker, you used the inbox to download existing data, allowing you to understand the details of the issues and update their status. Additionally, you collected new issues and updated their status as well.

Assignment 5: Use Field Maps for situational awareness and inspections in offline mode

Requirements:
You should base the homework for your Field Maps assignment on chapter 2.
1. Enable your web map for offline use.
2. Ensure your Geofence layer is also available offline.
3. Create at least one preplanned offline area.
4. Remove any calculation expressions that are not supported offline.
5. Create an on-demand offline area in the Field Maps app.
6. Copy the VTPK basemap of Citrus Plaza from https://arcg.is/XWSrP to your device.
7. Edit existing data and collect new data.
8. Share your layers and map with your instructor through a group.

What to submit:
- The URL or QR code linking to your web map.
- A screenshot of the Field Maps app showing your on-demand offline area.
- A screenshot of the Field Maps app showing a new feature collected on top of the side-loaded basemap.

Chapter 6
Workforce coordination and location sharing

Objectives
- Grasp the importance of workforce coordination.
- Recognize the advantages of location sharing.
- Set up assignment schemas and filters using Survey123 Connect.
- Work with assignments through the Survey123 app inbox.
- Enable tasks and configure contextual actions with Field Maps Designer.
- Navigate and operate tasks within the Field Maps app.
- Create track views.
- Share locations using mobile apps in the field.
- Monitor, visualize, and analyze location data as a manager.

Introduction
Mobile GIS operations are inherently collaborative, typically involving managers, dispatchers, and mobile workers who come together to effectively execute field tasks. Successful field operations hinge on robust operational management, which includes planning, organizing, and supervising the necessary work and determining who will carry it out. This chapter delves into the mechanisms of workforce coordination, starting with how assignments are facilitated by Survey123 Inbox filters and Field Maps Tasks. It also explores the advantages of location sharing for enhancing workforce coordination and its broader applications. The tutorials include three use cases: the first use case uses Survey123 Connect to set up the assignment schema and inbox filter, followed by experimenting with assignments using the Survey123 field app. The second use case enables tasks on a feature layer and configures contextual actions with Field Maps Designer, and then experiments with tasks using the Field Maps app. The third use case focuses on location sharing, including setting up a track view as an administrator, sharing locations using mobile apps, and visualizing track data through heat maps and time slider animations as a manager or dispatcher. The tutorial underscores the integration of technology and teamwork in Mobile GIS operations, providing readers with strategies for enhancing workforce coordination.

The importance of workforce coordination

As soon as there is more than one mobile worker and more than one field task, the need for workforce coordination becomes evident. Questions arise such as which worker should undertake which task. In larger deployments of field operations, where hundreds or even thousands of tasks and mobile workers intersect, the efficiency and effectiveness of coordinating among mobile workers, dispatchers, and managers become paramount.

The fundamental aspects of workforce coordination involve creating assignments, distributing tasks, notifying teams, and continuously monitoring their status. Mobile GIS requires a solution where tasks are assigned based on criteria such as task priority, worker availability, proximity to the task location, and worker expertise. This solution should enable managers or dispatchers to distribute workloads efficiently, ensuring the right person is assigned to the right task at the right time and providing the necessary data and tools to mobile workers.

Workforce coordination also necessitates timely communication between the office and the field. For instance, once a dispatcher assigns tasks, mobile workers should be able to view their assignments—or a spatial to-do list—on the map and in a list, potentially receiving push notifications. Mobile workers should be able to search and filter the tasks based on priority, distance, task type, and other attributes. Conversely, as workers update their progress, managers or dispatchers can instantly view these updates. This capability not only enhances transparency across the organization but also facilitates quick adjustments to plans as field conditions change.

Workforce coordination with Field Maps Tasks

Field Maps Tasks represents a new-generation coordination capability, effectively replacing the previous ArcGIS Workforce app. With Field Maps Tasks (figure 6.1), organizations can efficiently coordinate and push assignments to their teams in the field, communicating priorities and detailed work instructions. Mobile workers receive notifications for new tasks and can access a dynamic to-do list. This list is available in both connected and disconnected environments, detailing necessary tasks and their specific locations. Field Tasks is highly flexible, allowing organizations to customize data schemas, task actions, and integrations with other mobile apps such as Survey123, as well as with third-party business systems.

Field Maps offers several enhancements over Workforce, including the following:
- **All-in-one workforce management:** Streamlined workflows eliminate the need to open multiple apps, with simplified offline support and no requirement for copying data across apps.
- **A flexible data schema:** Instead of creating its own assignment layers, Field Maps enhances your existing layers, automatically adding task-related fields, including esritask_type, esritask_assignee, and esritask_status. these fields can be populated and edited using any app or programming language.

- **Context-driven actions:** Menus adjust based on task status, providing more relevant options.
- **Enhanced integration control:** Field Maps supports tighter integration with other apps and business systems through configurable forms that allow Arcade scripting.
- **Comprehensive task details:** View all task information, including instruction manuals in PDF format, directly in pop-ups.
- Task creation on the go: Mobile users, including workers and supervisors, can create tasks, pick up unassigned work, or take over tasks assigned to others.
- **Customizable task sorting and filtering:** Mobile workers can sort and filter tasks in various ways to better manage their workload.

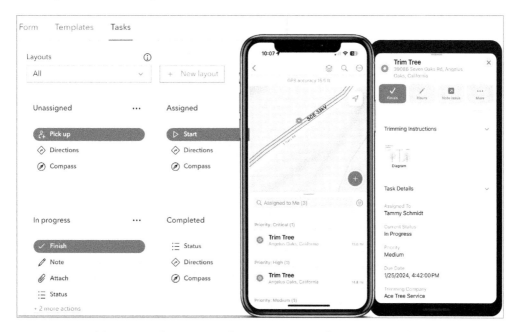

Figure 6.1. Field Maps Tasks replaces the previous Workforce product, laying out an assignment data schema and providing flexible context-driven actions.

Workforce coordination with the Survey123 inbox filter

In the previous chapter, you learned how the Survey123 inbox downloads all existing features into every mobile worker's inbox. Although this visibility helps workers understand the surrounding issues, it can lead to confusion and conflicts, as multiple workers might respond to the same issue. To avoid such conflicts, the Survey123 inbox enables survey creators to configure a query that filters features (figure 6.2), allowing each worker to see only the tasks assigned to them. Creators can also add fields such as status and assignment dates, taking advantage of the open and flexible schema.

Mobile workers can review their tasks in a list or on a map, displayed alongside their current location. This visualization aids them in sequencing their work based on task priorities and proximity. they can also update task statuses and communicate these changes back to the office, ensuring efficient workflow and coordination.

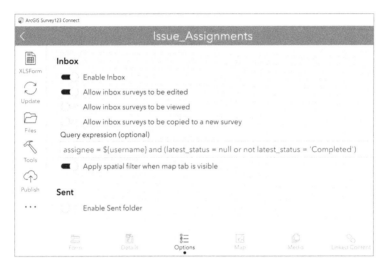

Figure 6.2. The Survey123 inbox query can filter assignments by username and assignment status.

Location sharing for enhanced situational awareness and collaboration

Location sharing is a crucial technology that connects organizations with their remote workers by providing real-time and historical data on user locations. This capability not only strengthens workforce coordination and collaboration but also enhances situational awareness and operational efficiency. The benefits of location sharing are manifold and applicable across various sectors.

- **Coordination based on proximity:** Knowing the precise location of each mobile worker allows organizations to optimize workflows by assigning tasks to those nearest to specific sites, reducing travel time and enhancing efficiency.
- **Improved situational awareness:** Essential during large-scale events, location sharing enables command centers to monitor the exact positions of responders in real time. This visibility ensures that personnel are always within reach, enhancing safety and enabling swift responses to emergencies or field changes.
- **Enhancing worker safety:** Continuous monitoring of field personnel's whereabouts allows organizations to rapidly respond to potential safety issues or emergencies, thereby reducing risks and ensuring worker safety.

- **Proof of work and audit trails:** Location sharing is invaluable for activities requiring strict compliance and verification, such as pipeline inspections or regulatory site visits. It facilitates the capture of detailed audit trails, ensuring inspections are thoroughly completed and compliance measures such as maintaining safe distances from assets are adhered to.
- **Determining area coverage and travel patterns:** Crucial for environmental or emergency response efforts—including homeless counts, invasive species removal, or search and rescue operations—location sharing ensures comprehensive area coverage. It aids in the effective deployment of resources and confirms that no critical areas are overlooked. Postevent analysis, including visualizing tracks through heat maps, time-enabled animations, and driving speed analysis, provides valuable spatial business intelligence.

Key components of location sharing

Deploying the location sharing capability within ArcGIS involves four essential components.

1. **Location sharing feature layer:** This is the central repository for all location data uploaded from mobile apps. It is composed of three distinct layers.
   - **Last known locations (LKL):** A point layer that holds a single record for each user, representing their most recently reported location.
   - **Tracks:** A point layer that records each location where a mobile worker has been tracked, forming a breadcrumb trail of their movements.
   - **Track lines:** A polyline layer available in ArcGIS Online, not available in ArcGIS Enterprise by default. It's a server-generated layer that connects and generalizes track points, allowing you to quickly visualize the paths traveled by each mobile worker.
2. **Track views:** As a subset of the location sharing feature layer, track views focus on the locations of a selected group of mobile users whose movements are being monitored, along with a designated group of users who are permitted to view these tracks. Tracks can be incorporated into maps, dashboards, and apps like any other feature layer, facilitating tailored visibility and analysis. You should never enable location sharing in more than one app on the same device simultaneously. Choose a single app to record location tracks.
3. **Field apps:** Mobile apps (figure 6.3), such as Field Maps, Survey123, and QuickCapture, enable the collection and sharing of location data. these apps upload tracks and LKLs to the location sharing layers. they are designed to function effectively in connected and disconnected environments, ensuring reliable data capture and sharing regardless of network availability. You should never enable location sharing in more than one app on the same device simultaneously. Choose a single app to record location tracks.

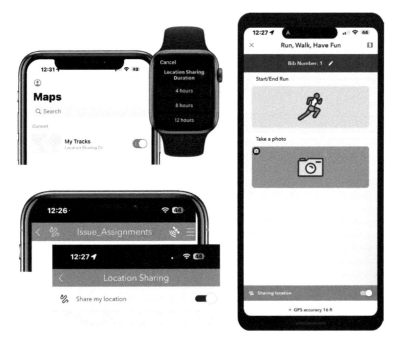

Figure 6.3. Mobile workers can share their locations using Field Maps and the companion Apple Watch app, Survey123, or QuickCapture.

4. **Track Viewer:** This web app (figure 6.4) allows track viewers to filter and examine accessible tracks. Use track views to observe both the last known location and the tracks of mobile users included in the track view. You can select a specific time span, choose individual mobile users, and even filter by GPS accuracy, among other options, to analyze the movement patterns of mobile users.

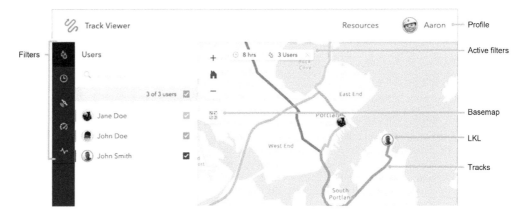

Figure 6.4. Track Viewer allows authorized users to filter and review tracks they have access to.

Tutorial 6: Coordinate workforce through assignments and location sharing

In this tutorial, you will experiment with the workflow of creating tasks, receiving assignments, enabling location sharing, and sharing your location, and then view the locations.

Use case 1: Dispatch facility maintenance workers to respond to campus issues using Survey123 (sections 6.1 and 6.2). This use case has the following functional requirements:
- Provide a web app for dispatchers to see the distribution and status of the reported issues.
- Enable dispatchers to assign the issues to mobile workers.
- Allow mobile workers to receive their assignments.
- Facilitate communication of assignment updates to the dispatchers.

Use case 2: Create assignments as a dispatcher and work on assignments as a mobile worker using Field Maps (sections 6.3 and 6.4, tutorials to be provided online).

Use case 3: Enable and use mobile worker real-time locations for workforce coordination and post analysis (sections 6.5 to 6.7).
- Configure whose locations will be shared and who can view the locations.
- Share locations using Field Maps, Survey123, and QuickCapture.
- View mobile worker real-time locations and play back the animation.

System requirements:
- Section 6.1 requires a Windows computer to run Survey123 Connect.
- Sections 6.5 to 6.7 require location sharing licenses unless you have a mobile user–type account.

6.1: Configure an assignment schema and inbox query using Survey123 Connect

In this section, you will learn how to create a new survey tailored for task assignment needs. This involves adding two new fields: one to store the assignees' usernames and another to store the most recent status of the issue. The updated survey is designed for use by mobile workers, specifically including campus facility maintenance responders.

1. On a Windows desktop, start Survey123 Connect and sign in with your ArcGIS Online account.

2. Download the Excel file from https://arcg.is/0y151T and drag it into Survey123 Connect. Click OK after the survey is created.

The new survey is called Issue_ Assignments and will display in preview mode.

3. On the left toolbar, click XLSForm.

   The Excel worksheet opens. The new survey is like the one created in section 5.5, with several enhancements that will be reviewed in the subsequent steps.

4. On the Survey tab, locate the assignee question.

   It's a text question and its read-only column is set to yes. This field will be used to store the usernames of the mobile workers assigned to the issues. The field is read-only on the form because its values are typically set by the dispatchers in the office using a web app, not by mobile workers in the field.

	A	B	C	I	K
1	type	name	label	readonly	calculation
16	text	assignee	Assignee	yes	
17	text	latest_status	Latest Status	yes	indexed-repeat(${status}, ${status_log}, count(${status_time}))
18	begin repeat	status_log	Status Log		
19	select_one list_status	status	Status		
20	dateTime	status_time	Status Time		
21	text	note	Note		
22	end repeat				

5. Locate the latest_status question.

   It's read-only, and its value is automatically calculated to be the value of last status in the status_log repeat using a formula.
   　The indexed-repeat() function is for extracting the value from a specific question in a repeat. It requires three parameters: the question name, the repeat name, and the repeat's index number. The following example returns the answer to the question status for the first record in the repeat status_log:
   　　　indexed-repeat(${status}, ${status_log}, 1)

   Because count(${status_time}) here returns the number of records in the repeat, the following formula returns the last value of the status question in the repeat.
   　　　indexed-repeat(${status}, ${status_log}, count(${status_time}))

6. In the Excel workbook, click the Settings tab.

The submission_url points to an existing feature layer, which has the additional attribute fields named assignee and latest_status, matching the schema of the form.

Next, you will enable the inbox and configure a query filter.

7. In Survey123 Connect on the lower toolbar, click Options.

8. For Inbox, turn on Enable Inbox.

9. For Query expression, type `assignee = ${username} and (latest_status = null or not latest_status = 'Completed')`.

   This expression filters the inbox to display only those records where the assignee field matches the username of the user currently using the Survey123 mobile app, and where the latest_status is not marked as completed. The functionality of this query will be demonstrated in the next section.

10. For Sent, turn off the Enable Sent folder.

11. On the left toolbar, click Publish.

    A message will appear to confirm that the survey is configured to use an existing feature layer, rather than creating a new one.

12. Click Publish Survey.

13. After publishing is completed, click OK.

6.2: Assign tasks and pick up tasks using Survey123

In this section, you will first take on the role of a dispatcher to assign some tasks to yourself, and then switch to the role of a mobile worker to pick up the tasks and update task status using the Survey123 mobile app.

Assignments or tasks can be assigned by directly editing the assignee field in the issues feature layer's table view, as illustrated, or by adding the layer to a web map and editing the assignee field within Map Viewer. This section will provide a dashboard as a sample dispatching web app.

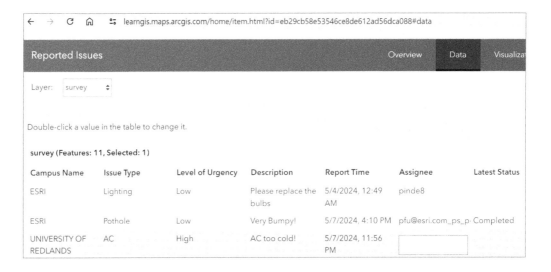

1. On your desktop, open a web browser and navigate to https://arcg.is/0PrunT0.

2. Review the elements of the Issue Dispatcher dashboard and read the instructions.

   This dashboard enables dispatchers to view reported issues both in a table and on a map. By observing the distribution and types of issues, dispatchers can more effectively determine which mobile workers are best suited to address specific problems.

3. Review the issues in the table.

   Some records have a latest_status whereas others—for example, those submitted using the surveys from earlier chapters—do not. Next, you will assign an issue to yourself.

4. Click an issue in the table that does not have a Completed status.

   The map zooms to the issue and displays its pop-up. A survey form displays allowing you to specify a username.

5. Type your username in the Assigned To text box.

   If you are signed in, your username should appear on the form, allowing you to copy and paste it.

6. Click Submit.

The table is configured to refresh every 30 seconds. Your name should be displayed as the assignee for the issue, unless it has been reassigned to another user in the meantime.

> **Note:** The dispatcher dashboard allows any user to assign issues to anyone if they know the ArcGIS Online username of the assignee. The possibility exists that your assignment may be overwritten by other users.

Next, you will switch to the mobile worker role.

7. On your mobile device, scan the QR code on the dashboard to launch the Survey123 mobile app and automatically open the inbox.

   Alternatively, you may use the survey you created in the previous section and tap the inbox.

8. Tap Refresh.

   You should see your assignment unless it's overwritten by others. The list of assignments is the result of the query filter you configured in the previous section.

9. If you don't see your assignment, it may be overwritten by other users. You may reassign or collect a new issue by scanning the QR code on the dashboard or using the survey you made in the previous section, refresh the dashboard, and then assign the new issue to yourself.

10. Tap the assignment in the inbox to open it. Scroll down to the status log section. Tap the Add button to add a repeat and set the status as Completed.

11. Tap Submit and tap Send Now.

12. Tap the inbox again to remove the assignment from your inbox. Tap Refresh and confirm that the assignment does not reappear, because its status is Completed and is excluded by the query filter configured.

13. Optionally, you may go back to test more assignments.

> **Working with tasks**
>
> Configure tasks using Field Maps Designer
> At the time of writing this book, the Tasks capability in Field Maps Designer has not been released. Once available, the tutorial will be provided online. Refer to this page for the tutorial: https://arcg.is/1rKf9H1.
>
> Assign tasks and pick up tasks using Field Maps
> Similar to the previous section, the tutorial will be provided online once the Tasks capability is available in the Field Maps mobile app. Refer to this page for the tutorial: https://arcg.is/0n0KTj.

6.3: Create a track view

This section has several prerequisites:
- An ArcGIS Online administrator role is required. You must be a member of the default administrator role in your organization to create track views.
- Location sharing must be enabled for your organization. If not, navigate to Organization > Settings > Organization Extensions and enable location sharing.
- Users to be tracked (those sharing locations) must have the add-on location sharing license.
- Users who will view the track views must have the privilege to view tracks.

1. On your desktop computer, start a browser, navigate to ArcGIS Online, and sign in with an administrator account.

2. Click the app launcher and click Track Viewer.

3. Click Create View.

4. Type Location Sharing Tutorial as the name of your view.

5. Click Create View.

   This will create a hosted feature layer view from the track layer for the organization.

   Next, you will add users—students in the class, for instance—to the track view.

6. On the Mobile Users tab, click the drop-down arrow to select a user and click Add. Repeat the process to add more users.

Chapter 6: Workforce coordination and location sharing   149

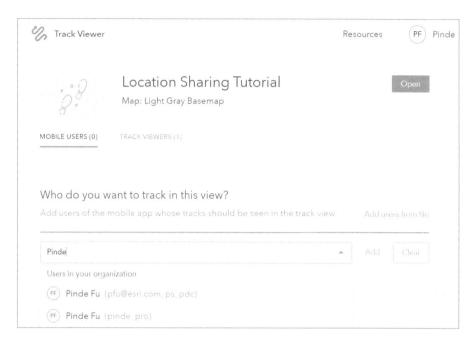

If there are many users to add, you can batch import them from a file or an existing group. Selected users need to have the location sharing license.

Next, you will configure who can access the track view.

7. Click the Track Viewers tab, click the drop-down arrow to select a user, and click Add. Repeat the process to add more users.

   This will create a group and add the selected users to the group. The above view is shared with this group. Selected users need to have the privilege to view location tracks.

8. Click Open to view tracks in Track Viewer.

   The track view is empty until users in the view share their locations.

   Users can share their locations using Field Maps, Survey123, or QuickCapture.

9. Enable your users to share their locations using one of the following apps:
   - **Field Maps:** No additional settings are required. Optionally, you can require location sharing for a specific web map.

- **Survey123:** From the app launcher, navigate to Survey123, click the Organization tab, click Settings, click Location Sharing, check Enable Location Sharing, and under Data, click Tracks and Last Known Locations. Click Save.
- **QuickCapture:** Select an existing QuickCapture project or create a new one. On the upper toolbar, click the Additional Settings button (three dots), click Location Sharing, and check Enable Location Sharing. For Last Known Locations, change the Update interval to every 1 minute. For Tracks, set the Upload interval to 10 minutes. Click Save.

6.4: Share your locations using mobile apps

In this section, you will assume the role of a mobile worker to share your locations, using Field Maps, Survey123, or QuickCapture.

1. If you use Field Maps:
    a. Open Field Maps on your device and sign in.
    b. In the Maps list, turn on location sharing in My Tracks.

    c. Set the duration—for example, 4 hours or until switched off.
    d. If prompted to allow Field Maps access to your location, tap Allow While Using the App.
    e. If prompted to allow Field Maps access to your motion fitness and activity, tap OK.

2. If you use Survey123:
    a. Open Survey123 on your device and sign in.
    b. Tap the Location Sharing button, as pictured.
    c. Switch on Sharing My Location.
    d. If prompted, allow Survey123 to access your location, motion, and activity.

Location Sharing Button

3. If you use QuickCapture:
   a. Open QuickCapture on your device and sign in.
   b. Open the QuickCapture project that has location sharing enabled.
   c. In the Start Location Sharing pop-up, click Start.

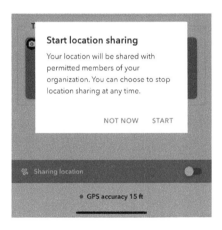

4. To see your tracks live on your device:
   - In Field Maps, tap My Tracks.
   - In Survey123, tap Location Sharing.

5. Allow location sharing to run for 15 minutes or longer. If possible, take a walk outdoors.

   Walking outdoors can achieve better location accuracy.

6.5: Monitor, visualize, and analyze locations shared

In this section, you will take on the role of a dispatcher or manager to review both live and stored locations. Monitoring the live locations of mobile workers enables dispatchers to identify who is nearest to a specific issue, facilitating more informed decision-making regarding task assignments.

1. On a desktop computer, start a web browser, navigate to ArcGIS Online, and sign in.

2. Click the app launcher and choose Track Viewer.

3. Click the track view created in the previous section, Location Sharing Tutorial.

4. In the Track Viewer, on the Users tab, select some or all users to display their tracks.

   Each user's last known location is displayed with a label of their initials.

5. Click a track point to see the speed, direction, battery percentage, and location accuracy in the pop-up.

6. Click the Time Span tab to see that you can filter the tracks by time.

7. Click the Activity tab to see that you can filter the tracks by activities, such as driving, walking, and stationary.

   The tutorial is designed to be straightforward. Track layers can be added to the same web map as the issues and visualized on the dispatcher dashboard. This setup enables dispatchers to effectively determine the most efficient assignments.

   Next, you will review the data collected in the data table view.

8. In the web browser, navigate to ArcGIS Online and sign in.

9. Click Groups.

10. In the list of groups, find the group for the mobile view you are in. Click View Details.

    The group name may be Location Sharing Tutorial, or other names as specified earlier.

11. Click the Content tab of the group.

By default, this group should contain only the feature layer view created as in the previous section.

12. Click the Options button (three dots) of the feature layer and click View Details.

13. Click the Data tab.

14. Click the Layer drop-down arrow and notice the three layers: Tracks, Last Known Locations, and Track Lines.

    The Application ID field in the Tracks and Last Known Locations layers shows which app the location was collected with.

    Next, you will play back the track points over time.

15. Click the Overview tab.

16. Click Open in Map Viewer.

17. In the Map Viewer layer list, expand the location sharing layer. Click the Tracks layer to select it.

18. Zoom in and out of the map until the track points appear.

19. On the Settings toolbar, click Properties.

20. In the Properties pane, under Time, turn on Enable Time.

21. On the Content toolbar, click Map Properties and apply the following settings:
    a. Under Time, click Time Slider Options.
    b. Set the Start and End Date and Time to a range where locations are collected.
    c. For Time Intervals, choose Total Time Divided into Equal Steps and set the count to 20.
    d. For Play rate, choose Fast.

22. On the map, click Play on the time slider and watch the locations animate.

    Next, you will visualize the locations in a heat map.

23. On the Settings toolbar, click Styles.

24. Under Pick a Style, click Heat Map.

25. Click Done.

    This heat map displays the most frequently visited locations by your mobile workers. This allows you to understand the areas most visited by your mobile workforce.

26. Save your web map.

In this tutorial, you explored the workflow for creating and managing assignments, enabling location sharing, and monitoring real-time locations to improve workforce coordination.

The Survey123 tutorial enhanced the data schema to support assignee names and the latest task status, enabled the inbox feature, and configured a query filter based on username and issue status. You then took the role of a dispatcher using a dashboard to review reported issues and assign tasks. Subsequently, you took the mobile worker role to view your assignments in your Survey123 inbox, update their status, and relay this information back to the dispatcher dashboard.

In the Field Maps tutorial, you enabled tasks on your feature layer and set up quick actions for these tasks using Field Maps Designer. Like the Survey123 approach, you took on the dispatcher role to assign tasks to yourself, and as a mobile worker, you collected and updated task statuses using the Field Maps app.

The location sharing tutorial experimented with the process for creating a track view as an administrator, specifying who can share and view locations. It covered sharing locations as a mobile worker and monitoring live locations as a dispatcher or manager, including playback of tracks and analysis of track patterns.

These tutorials are designed for clarity and ease of use. Track layers can be integrated into the same web map as the assignments and visualized on the dispatcher dashboard. This visualization allows dispatchers to quickly understand spatial relationships and facilitate effective task assignments. Although you used the dashboard to assign tasks in these tutorials, detailed discussions on dashboard design and implementation are not included in this chapter. For more detailed instructions, please refer to chapters 7 and 8, as well as other books and online resources.

Assignment 6: Create a survey or web map to support workforce coordination using assignments or tasks

Requirements:
- Create a new feature layer. Do not reuse the campus issues or hydrants layers from the tutorials. Related tables are not required.
- A dispatcher dashboard or other web app is not required. Allow your instructor to assign tasks by editing the assignees in the table view of the assignment feature layer.
- You may use the Survey123 inbox approach or Field Maps Tasks.
    - If using the Survey123 inbox, a query filter with the username is required.
    - If using Field Maps Tasks, provide six appropriate quick actions for each status.
- Share your content items publicly or through a group so that your instructor can access them.

What to submit:
- The QR code or URL to your survey (if using Survey123) or to your web map (if using Field Maps).
- The URL to your assignment feature layer, allowing your instructor to assign tasks by changing the assignees in the table view.

Chapter 7
Responsive web apps for mobile devices

Objectives
- Consider the value of responsive web apps in Mobile GIS workflows.
- Learn the considerations for building mobile browser-based apps.
- Use the basic workflow to configure ArcGIS Dashboards web apps.
- Configure mobile views of dashboards.
- Use the basic workflow to configure ArcGIS Experience Builder web apps.
- Optimize Experience Builder app layouts for mobile devices.

Introduction
The previous chapters have discussed native mobile apps, associated designers, and their workflows. This chapter shifts focus to browser-based web apps. Although these apps typically do not work in offline conditions, they perform well when connected to the internet. they provide an effective option for delivering data to the field for situational awareness and streamlining data collection and editing. Additionally, browser-based apps play a crucial role in office-based quality control, data review, editing, approval, and assignment processes. This chapter will cover the range of ArcGIS configurable web apps, including ArcGIS Instant Apps, ArcGIS StoryMaps℠, Experience Builder, Dashboards, and ArcGIS Hub℠. We will explore their capabilities, compare their features, and discuss their support for responsive web design. The tutorial section begins by creating and exploring a StoryMaps story on your device. Next, it dives into using a dashboard and Experience Builder for in-office data review on desktop browsers. Further, the tutorial adapts these apps for mobile use by optimizing the layouts and simplifying the functions, supporting effective use in the field when connected.

ArcGIS configurable responsive web app templates and app builders

Browser-based web apps generally require an internet connection to function adequately, but when connected with sufficient bandwidth, they can effectively deliver and collect data from the field. Responsive web apps are designed to adjust their layouts to accommodate the smaller screen sizes of mobile devices, ensuring a smooth user experience for field operations. Additionally, web apps are commonly used for data review in office settings using desktop browsers. This versatility makes web apps a crucial component of the Mobile GIS workflow and an important option for field use.

ArcGIS provides a suite of configurable web app templates and app builders (figure 7.1), including Instant Apps, StoryMaps, Dashboards, Experience Builder, and Hub. Web apps created using these tools are responsive across various screen sizes.

- **Instant Apps**: Enables users to quickly design and deploy web apps tailored to specific tasks and audiences. these template-driven tools require minimal configuration, making it easier for users to visualize, present, and share geographic information. Instant Apps are ideal for projects requiring simple functions, rapid deployment, and straightforward user interfaces.
- **ArcGIS StoryMaps**: Allows users to craft interactive stories by combining maps with narrative text, images, and multimedia content. This tool is designed to engage and inspire audiences with geographically enriched storytelling, making it perfect for educational purposes, advocacy, and public engagement. ArcGIS StoryMaps provides a user-friendly interface that helps you connect with your audience through compelling geographic narratives.
- **Dashboards**: Displays geographic information as interactive maps, charts, gauges, and more, allowing for real-time data visualization and awareness at a glance. Dashboards are widely used in emergency management, public services, and business operations to provide insights.
- **Experience Builder**: Offers a flexible and advanced approach to creating web apps with rich data and rich user experiences. With drag-and-drop functionality, users can integrate maps, text, images, and other widgets into fully responsive layouts that work across all devices. Experience Builder is designed for those who want to customize and extend their web applications beyond standard templates.
- **Hub**: An engagement platform that enables organizations to connect with their communities through open data, web apps, and interactive initiatives. It supports collaboration between government, the private sector, and the public to tackle community challenges effectively. By facilitating easier access to data and tools, Hub helps foster a more informed and involved community, advancing public awareness and participation.

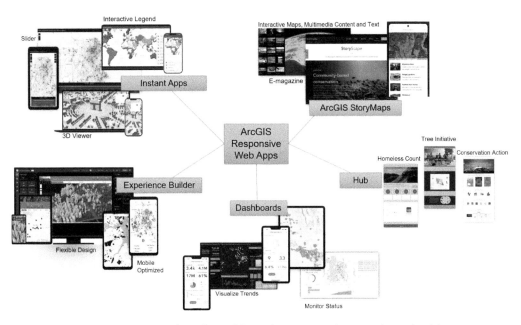

Figure 7.1. The ArcGIS suite of configurable web app templates and app builders, including Instant Apps, ArcGIS StoryMaps, Dashboards, Experience Builder, and Hub, can create responsive web apps with functionalities for data collection, review, and editing, both in the office and in the field.

Basic workflow to create dashboards

Dashboards are composed of configurable widgets, such as maps, lists, charts, gauges, indicators, and tables. these widgets are called elements or components. Most of these elements are data-driven, representing the information you want to present to your intended audience. Dashboard apps are commonly designed to display multiple visualizations that work together on a single screen. With editing-enabled tables and embedded components, such as Survey123 forms, dashboards can support data editing. A dashboard can be configured with both desktop and mobile views (figure 7.2), each tailored with different elements and layouts to meet the specific needs of the device. When users open the dashboard, they automatically see the optimal view for their device, ensuring a seamless and effective experience across platforms. The basic workflow to create a dashboard generally includes the following steps:

1. Prepare data layers and web maps or scenes.
2. Create a dashboard and add elements.
3. For each element, configure its data source, style, and layout.
4. Set up interactivity between the elements by configuring actions and targets.
5. Refine the layout.
6. Save, preview, and share.

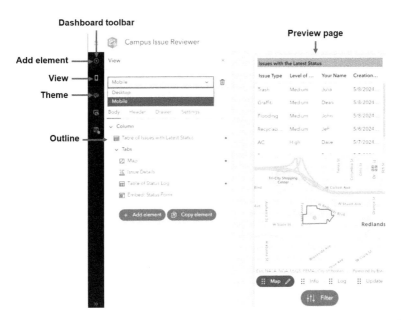

Figure 7.2. The Dashboards app allows users to create responsive web apps by configuring separate desktop and mobile views, each with optimal layouts and functions.

Basic workflow to create web experiences

In ArcGIS technology, the web apps created using Experience Builder are called web experiences. Experience Builder has the following workflow to create web experiences:

1. **Pick a premade template or start from scratch:** A template includes a collection of preconfigured widgets arranged in certain layouts and styles.
2. **Select a theme:** A theme is a preset style scheme for the appearance of your app.
3. **Add source data:** Bring in 2D web maps, 3D web scenes, and feature layers created by you or shared with you.
4. **Add and configure widgets:** After widgets are added, you can configure their data sources, styles, and actions.
5. **Refine layouts for all devices:** Optimize different page layouts and unique designs (figure 7.3) for large, medium, and small screen sizes.
6. **Save and share:** Save, preview, publish, and share.

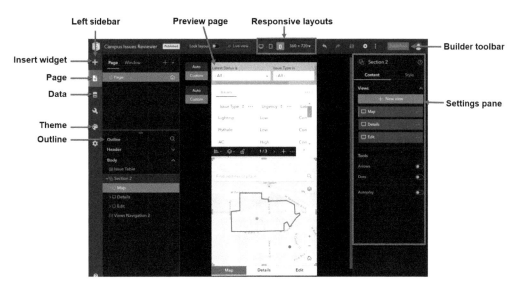

Figure 7.3. Experience Builder provides a what-you-see-is-what-you-get environment for creating mobile phone web apps with flexible layouts, rich user experiences, and versatile data visualization and editing capabilities.

Sources, targets, and actions

Experience Builder and Dashboards incorporate the concept of actions, which enable widgets to interact and work together seamlessly. Configuring actions involves three key components.

- **Sources (also known as triggers):** these events are generated by the source widget, such as map extent changes or record selection changes.
- **Targets:** these elements perform actions in response to the trigger. A target can be a specific widget or the framework data. When the framework is chosen as the target, the action is performed globally, affecting all relevant widgets and pages within the app.
- **Actions:** these specific business logic tasks are executed by the targets, such as panning and zooming on a map. Some actions require additional configuration. For instance, when configuring a filter action based on different data, you can filter the target data by configuring an attribute or spatial relationship.

Considerations for mobile web apps

When creating web apps for mobile devices, a set of design considerations must be prioritized to ensure usability, performance, and overall user satisfaction.

- **Limit the number of elements or widgets:** On mobile devices, screen real estate is limited. It's crucial to keep the user interface clean and uncluttered. Limiting the number of on-screen elements and widgets not only helps in making the app look more organized but also improves app performance by reducing memory and processing requirements.
- **Optimize the layout:** Layouts should be designed with mobile ergonomics in mind. This means big, easily tappable buttons and generous spacing for touch targets. Navigation should be intuitive, often at bottom for easy figure access even with one hand. To save space, you can stack elements in Dashboards or use sections and widget controllers to stack widgets in Experience Builder.
- **Stack multiple widgets:** This is often needed to save space.
- **Minimize text:** Excessive text can overwhelm mobile users, who typically prefer quick interactions. Use concise language and bullet points or icons. Essential information should be front and center, whereas additional details can be made available through expandable sections or additional tabs.
- **Consider mobile limitations:** Mobile devices do not support hover state interactions, which are common in desktop environments, well. Instead, implement design interactions that are touch-friendly, such as taps or swipes. Also, consider the implications of virtual keyboards appearing on the screen and ensure that forms and important content are not obscured when the keyboard is active.
- **Create simple visualizations:** Complex graphs or charts can be hard to read on smaller screens. Simplify visualizations to include only the most relevant data, using bold colors and clear, legible fonts. Interactive elements should be large enough to manipulate without accidental inputs.
- **Build mobile-friendly maps:** Limit the number of operational layers to ensure smooth performance and enhance readability. Symbology must be clear and must contrast well with the basemap to be easily discernible on small screens. Also, consider the use of GPS features to enhance map interactivity, such as centering the map on the user's current location.

Tutorial 7: Explore and configure responsive web apps for data review

In this tutorial, you will first explore ArcGIS StoryMaps stories to see how they respond to mobile screens. Then, you will focus on Dashboards and Experience Builder. You will explore the desktop views of a dashboard and a web experience, and then configure these apps by adding mobile views to them. The resulting web apps can support both in-office and in-field data review.

Data:
You are provided with the following layers, maps, and apps:
- The campus issues layer and its related status log table, created in section 6.2.
- A web map that includes the above layers, table, and a campus polygon layer.
- A Survey123 form for editing issues.
- A dashboard with a desktop view for in-office data review.
- A web experience with a desktop layout for in-office data review.

Requirements:
- Copy the provided dashboard and web experience.
- Explore the copies to learn how the desktop views are configured.
- Modify the copies to create mobile views.
- The resulting web apps should be responsive to phone screen sizes and support data editing in the field.

7.1: Explore ArcGIS StoryMaps stories on your mobile devices

Although not the primary focus of this book, ArcGIS StoryMaps stories have gained popularity as an engaging method to present data, including those collected from the field. With millions of stories already created, this section provides an opportunity for you to select and explore several of these compelling visual narratives.

1. On your smartphone or tablet, navigate to the ArcGIS StoryMaps Gallery at https://arcg.is/1Lmyq50.

2. Browse the gallery and test at least three stories.

3. Change the screen orientation to see how the stories respond.

   While reviewing the story maps, reflect on how their elements, capabilities, and interactive features can be incorporated into your Mobile GIS workflows or projects.

7.2: Explore the desktop view of a dashboard

This section will begin by guiding you through the process of copying a dashboard. Since you are the owner of the copy, you can then explore it in editing mode to learn about its configuration.

1. Open a web browser and navigate to https://arcg.is/KKnDL2.

2. Sign in to your ArcGIS Online account.

   The short URL is expanded into its original, longer form. The Create New Dashboard page displays. The URL has the code `dashboards/new#id=`, which is the syntax to create a new dashboard by copying an existing dashboard.

3. Click Create Dashboard.

   The dashboard will be copied to your content. The editing mode of the dashboard appears.

Next, you will explore the functions of the dashboard, which has the following components:
- A header with selectors of issue status and issue type
- An Issue table, which is a table widget
- Number of Issues, which is an indicator widget
- An Issues by Category column chart, which is a serial chart widget
- Tabs to show issue details and the status log and to update issue status

4. In the header, note that making any selections in the issue status and type selectors will filter the number, chart, and table widgets accordingly.

5. Hover over the Select an Issue Status selector, hover over the Options button (three dots), and click Configure.

The Category selector pane opens. On the Data tab, this widget uses the Issues layer and the Latest Status field.

6. On the left, click the Actions tab and click Filter to expand the settings.

The widget filters the Indicator number of issues, the map, the chart, and the table. This is why when you selected an issue status, the number, map, chart, and table updated accordingly.

7. In the lower right of the page, click Cancel.

Next, you will explore the interaction from the Issue table to the Map, Details, and Status Log widgets.

8. In the Issue table, click an issue to zoom to the issue location. The issue flashes on the map and the details are shown in the Details widget. Click the Status Log tab so that the issue's status history displays.

9. Hover over the table, hover over the Options button, and click Configure.

The Table pane appears. On the Data tab, this widget uses the Issues layer and has several attribute fields selected for display.

10. Click the Actions tab. Click Flash, Pan, and Zoom to expand the settings. The Map widget is selected for these actions.

11. Click Filter and note the following filter settings:
    - It filters the Details widget.
    - It filters the Status Log list.
    - It filters the Embedded Survey widget, with GlobalID as the Source field in the issues layer and ParentGlobalID as the Target field in the Status Log table.

12. Click Cancel.

Next, you will examine how the embedded Review form is configured.

13. At the bottom of the tabs, click the Review tab.

A Survey123 form for updating issue status appears. The form is designed to add records to the Status Log table.

14. Hover over the title of the Update Status Form widget, hover over the Options button, and click Configure.

The Embedded Content pane appears. Its data type is Features, its content type is Document, and its URL is structured as follows:

https://survey123.arcgis.com/share/321bdf5716564d1f996c14ae944fffe9?hide=header,footer,theme,leaveDialog,description&field:parentglobalid={field/globalid}

The URL points to a Survey123 form identified by the item id. The URL hides the survey's header, footer, theme, "leave" dialog box, and description. Moreover, the URL passes the GlobalID of the issue selected to the globalid field of the Issues layer. This is to ensure that the correct record is edited.

15. Click Cancel to exit the Settings pane.

In this section, you copied a dashboard and studied how it is configured.

7.3: Configure a mobile view for the dashboard

In this section, you will enhance the dashboard you copied in the previous section by adding a mobile view, as illustrated in figure 7.2. This addition will enable the dashboard to be effectively used for reviewing data on smartphones.

Before starting the tutorial steps, it is helpful to review figure 7.2 to understand the mobile view that you will be creating.

1. If it isn't open already, open the Edit Dashboard mode.

   Alternatively, you can locate this dashboard in your content, open its item page, and click Edit Dashboard.

2. On the dashboard toolbar on the left, click the View button.

3. Click the Add Mobile View button.

   The mobile view of the app is empty by default. Next, you will copy some elements from the desktop view.

4. Under Body, click the Copy Element button.

5. Select the table to copy it.

6. Repeat the previous two steps to copy the map, the Status Log list, and the embedded survey.

   With these widgets, the dashboard might appear crowded. Next, you will organize some of the widgets by stacking them.

7. In the Mobile view, hover over the Map widget, hover over the Options button, press and hold the Drag Item button, and move it toward the center position indicator of the Status Log List widget. Release the mouse button when the text changes to Stack the Items.

Your stack should have the List, Embedded Content, and Map tabs.

Next, you will add the Details widget to the stack.

8. On the dashboard toolbar, click the Add Element button.

9. Hover over the stack and click the Add button in the center.

10. Select the Details widget.

11. In the Select a Layer window, select the Issues layer.

12. For Maximum Features Displayed, choose 1.

13. Click the General tab and set its name to Issue Details.

14. Click Done and then click Save to save your dashboard.

Throughout the section, save your dashboard frequently to prevent any loss of your work.

Next, you will resize the stack, rename the tabs, and rearrange their order.

15. If the stack is at the top of the view, drag it to the bottom, below the Issues with Latest Status table.

Keeping the tabs at the bottom of the page will make it easier to operate with one hand because it brings them within easier reach of your fingers.

16. Hover over the border between the stack and the table above it until the cursor changes to cross hairs. Drag the border to resize the table to occupy about 35% vertical space of the page.

17. Double-click the Embedded Content tab and rename it Update.

18. Similarly, rename the Details tab Info and the List tab Log.

19. Drag the tabs to reorder them as follows: Map, Info, Log, and Update.

   Next, you will reduce the number of fields in the table to make it more readable.

20. Hover over the table, hover over the Options button, and click Configure.

21. Under Value Fields, click X to remove the fields and keep only Issue Type, Level of Urgency, and Latest Status.

   Note: When you copy elements from the desktop view, actions are not preserved.

   Next, you will configure the actions of the table to display the issue selected on the map and filter the information across the Info, Log, and Update tabs.

22. On the Actions tab, click Flash and turn on the map.

23. Click Pan and Zoom. Turn on both actions for the map.

24. Click Filter and apply the following settings:
    a. Turn on Embed Survey and click Render Only When Filtered.
    b. Turn on Issue Details and click Render Only When Filtered.
    c. Turn on the Status Log list and click Render Only When Filtered. For Source field, click GlobalID, and for Target field, click ParentGlobalID. these are the key fields of the relationship between the Issues layer and the Status Log table.
    d. Click Done.

25. Click Save.

    Next, you will test the mobile view you just configured.

26. Click a record in the Issue table and explore each of the tabs—Map, Info, and Update—to see how they have all updated to reflect the record you selected.

27. In the upper left of the page, next to the Issue Reviewer title, click the menu button (three lines) and click Dashboard Item Details.

28. Click Open Dashboard.

29. Copy the URL of the page and send it to your phone—through email, for instance.

30. Open the dashboard on your phone, sign in to your ArcGIS Online account, and explore the functions and responsiveness of the mobile view.

    > **Note:** Dashboards respond only to screen sizes, not to browser sizes. Therefore, reducing the browser size on your desktop computer will not trigger the mobile view because the actual screen size is constant and is large. You may scan the QR code below with your phone or navigate to https://arcg.is/1zqjni0 to explore the mobile view configured in this section.

In this section, you configured a mobile view of a dashboard that enables reviewers to review issues and update their status on their smartphones.

7.4: Explore the desktop layout of a web experience

This section will guide you through copying an Experience Builder app. Since the copied experience will be yours, you can explore it in the editing mode to understand how the web experience is configured. For simplicity, the functionality of this experience is kept like that of the dashboard used in section 7.2.

This section provides a predesigned web experience template. You will use this template to create a new app, effectively creating a copy of the web experience.

1. In a web browser, go to https://arcg.is/1zm80b to sign in to ArcGIS Online.

   Alternatively, you can go to ArcGIS Online, sign in, and search for campus reviewer template owner:GTKMobileGIS. You will need to deselect the Only Search in <your ArcGIS Online organization> filter and open the item page of the found template.

   This is a predesigned Experience Builder app template with a desktop layout designed for reviewing data in the office, typically on a large screen.

2. Click Create Web Experience.

   This creates an app, essentially copying the template app.

3. Change the item title to Campus Issue Reviewer (your name).

   Next, you will explore the app, similar to what you did in section 7.2.

4. On the builder toolbar, click Publish and then click Preview.

   The web experience opens in a new Browser tab. Next, you will explore its functions.

5. Note the following widgets:
   - Number of Issues, which is a text widget referencing dynamic content from the Issues layer.
   - Issues by Type column chart, which is a chart widget.
   - Issue Table widget.
   - The tabs, which are a Section widget with three views and a Views Navigation widget. The three views are the Info tab, which has the Feature Info widget; the Edit Form tab, which has a Survey123 widget; and the Edit Widget tab, which has the Edit widget of Experience Builder.

6. In the header, observe the Status and Issue Type filter.

   Making any selections in these two selectors will filter the number, chart, and table widgets mentioned earlier.

7. Click an issue in the Issue table.

   The map zooms to the issue location and highlights the issue. The details of the issue display on the Info tab.

   Next, you will explore the different ways to edit the data using the Survey123 widget and the Edit widget.

8. Click the Edit Form tab.

   A Survey123 form displays.

9. On the form, scroll down to the repeat. Click the Add button to add a new status. Set the status as In Progress and click Submit.

10. In the Issue table, click Refresh to update the status of the issue.

11. Click the Edit widget.

    The Edit widget appears, configured to edit only attributes in this space, although it is capable of editing geometries as well.

12. In the Edit widget, update the level of urgency and then click Update.

13. In the Issue table, click Refresh.

    The Level of Urgency of the issue has been updated.

    Next, you will explore how these functions are configured.

14. Click the previous Browser tab to return to the configuration mode of the experience.

15. On the toolbar on the left, or left sidebar, click the Data button to display the Campus Issue Reviewer map in the app.

The web map has a Source button for you to review the web map's item page with details if needed.

16. On the left sidebar, click the Page button.

This app has only one page. Its outline displays the header and the body, including all the widgets used in the app.

17. On the Preview page, hover over the header and click Edit Header.

18. Click the Status and Issue Type filter.

Its settings appear on the right side. It is configured to use the Issues layer.

19. Click the Issues filter.

20. Click SQL Expression Builder and review how the filter is configured on the Status and Issue Type fields.

Because it filters the Issues layer, the filter affects all the widgets using this layer, including the number of issues, the chart, the table, and the map.

21. Click Cancel.

22. In the Page view, click the Issue table to open the Issue Table settings. Click Action.

Under Message Action, when a record is clicked or selected in the table, the map pans to, zooms to, and flashes the selected record.

Clicking a record in the table selects the record. The selected record is passed to the Feature Info, Survey, and Edit widgets.

Next, you will explore how the Edit Form section is configured.

23. In the Page outline, expand Section, expand the Edit Form view, and click the Survey widget.

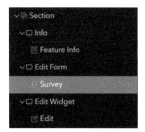

In the Survey widget settings, the mode is set to Edit an Existing Record. Its source layer is set to the Issues layer, which means that this widget is configured to receive data from the layer when an issue is selected.

24. If you made any changes, click Save and Publish. Optionally, share the web experience with everyone or a group.

In this section, you explored the functions of the experience and how the different widgets interact with one another using actions.

7.5: Configure a phone layout for the web experience

In this section, you will enhance the web app you created in the previous section by adding a phone layout, as illustrated in figure 7.3. This addition will enable the experience to be effectively used for reviewing data on mobile devices.

Before starting the tutorial steps, it is helpful to review figure 7.3 to understand the mobile layout that you will be creating.

> **Note:** When designing the phone layout, do not delete any widgets or screen groups, even if they are not needed for the phone layout. Deleting them would also remove them from the desktop layout. Instead, move them to the pending list.

1. On the builder toolbar, click the Edit the Page for Small Screen Devices button (small vertical rectangle).

   By default, the header and body of the app are adjusted using Auto Layout, where Experience Builder automatically makes changes. Many widgets in the app are now too small to read and operate. You'll customize the header first.

2. For the header, click Custom.

   A small dialog box appears, asking whether you want to enable it.

3. Click OK.

4. Hover over the header and click Edit Header.

5. Click the title. On the widget toolbar, click the Move to the Pending List button.

   Now, the title has been removed from the phone layout while it is still available in the desktop layout.

6. Drag the Status and Issue Type filter to the left. Resize its width to fit in the header.

Next, you will customize the body layout.

7. For the body, click Custom.

   A small dialog box appears, asking whether you want to enable it.

8. Click OK.

   Next, you will move the chart to the pending list so that it is removed from the phone layout to make space for other widgets.

9. Click the chart widget and click Move to the Pending List.

10. Perform the same for the Issues by Type label.

    Next, you will duplicate the Section widget, the map, and the Views Navigation widget. You will move the original copies to the pending list and work with the copies.

    Views in a section are not convenient to select on the page. Selecting them in the outline is easier.

11. If the Page outline is not open, click the Page button on the left sidebar to open the Page outline.

12. Under Outline, click the Section widget, click the More button (three dots), and click Duplicate.

    Section 2 is created and displayed in the outline.

13. Under Outline, click Section again, click the More button, and click Move to the Pending List.

    Next, do the same for the Map widget and the Views Navigation widget.

14. Under Outline, duplicate the Map and Views Navigation widgets. Move the original copies to the pending list.

    Next, you will remove the Edit Widget view and add a new view to display the map.

15. Under Outline, click Section 2 to select it.

16. In the Settings pane on the right, under Content, perform the following in the Views section:
    a. Click Edit Widget 2, click the More button, and click Delete.
    b. Click Edit Form 2, click the More button, click Rename, and change its name to Edit.
    c. Click Info 2, click the More button, and rename it Details.
    d. Click New View.
    e. Click the new view created, click the More button, and rename it Map.
    f. Click the Map view. While the Map view is selected, on the Preview page, drag the existing map (Map 2) to the Map view. Resize the map to occupy the full height and width of the view.

17. In the outline, click Section 2 again to select it.

18. In the Settings pane, under Content, drag the views to set their order, from top to bottom, to Map, Details, and Edit.

Now the body of the page contains three main elements: the Issue table, Section 2, and Views Navigation 2. Next, you will move and resize them to have a layout like that in figure 7.3.

19. Under Outline, click Views Navigation 2 to select it. On the Preview page, drag it to the bottom of the page. Resize it to take up the full width of the page.

20. Under Outline, click the Issue Table widget to select it. On the Preview page, drag and resize it to occupy the top 40% of the height and the full width of the page.

21. Under Outline, click Section 2 to select it. On the Preview page, drag and resize it to occupy the remaining space between the table and the navigation bar.

22. On the builder toolbar, click Save and then click Publish.

23. On the builder toolbar, click Preview.

    The app will open in a new Browser tab displaying the desktop view.

24. Resize your browser window to simulate a phone size. Notice that the phone view appears.

25. Share the app with the public and test the app on your phone.

    You can test the app on your phone by navigating to https://arcg.is/1qGrCT or scanning the QR code provided.

The ArcGIS suite of configurable web app templates and app builders offers responsive user experiences and sophisticated functionalities. Each product requires time to fully understand and master. This tutorial provided a quick and basic introduction to Experience Builder and Dashboards. You explored the desktop views and configured mobile views. The desktop views allow for data review in the office on larger screens, whereas the mobile views facilitate viewing, collecting, and editing data in the field. The tutorials demonstrated the value of responsive web apps for field awareness and data capture. Experience Builder, Dashboards, Hub, Instant Apps, and ArcGIS StoryMaps offer even more capabilities to explore and learn.

Assignment 7: Build a responsive web app for data review and editing

Requirements:
1. Choose either Experience Builder or Dashboards.
2. The web app must offer different layouts for the desktop and mobile phones. Tablet layout is not required.
3. Enable viewing existing data on maps and in a list or table.
4. Enable editing data through an edit widget or a survey form.
5. Configure actions: Clicking on a record within the list or table should trigger zooming/panning of the map to the selected item and opening the record for editing.
6. Use your own feature layer or layers and one or more web maps. You may reuse previously created layers and maps.
7. Share your layers, maps, and app with everyone or with your instructor through a group.

What to submit:
- The URL to your web app
- The URL(s) to your feature layer(s)
- The URL to your web map(s)

Chapter 8
Integration with enterprise systems

Objectives
- Explain webhooks.
- Explain the ArcGIS connectors in Microsoft Power Automate.
- Design Survey123 feature report templates and learn the expression syntax.
- Configure flows in Power Automate.
- Integrate mobile workflows with Microsoft Outlook, OneDrive, and Teams.

Introduction
Integration stands as a critical aspect among the four main patterns of Mobile GIS applications: data capture, field awareness, planning and coordination, and integration itself. It ensures that data and insights gathered in the field are not isolated. This process allows Mobile GIS to directly feed data into the enterprise GIS, where it can be analyzed, visualized, and shared across the organization in real time. In the previous chapter, you designed Experience Builder and Dashboards web apps to visualize, chart, review, and edit the data collected in the field. ArcGIS technology provides a Python API and webhook connectors to further help users in automating integrations, manual tasks, generating reports, and streamlining field workflows. This chapter will explore how to integrate Mobile GIS with widely used collaboration and productivity tools, such as Outlook, OneDrive, and Teams. Through these technologies and methods, Mobile GIS becomes more powerful, extending far beyond simple data collection by automating manual tasks, facilitating communication, and fostering more informed decision-making across the enterprise.

Integrating Mobile GIS with enterprise systems
Integrating Mobile GIS with enterprise systems facilitates real-time data access and sharing. This immediacy ensures that all stakeholders, regardless of location, have access to the most current information, promoting better-informed decisions. It fosters a collaborative environment by breaking down silos between departments. Enhanced data sharing and communication capabilities enable teams to work together more effectively, streamlining workflows and increasing productivity.

Integrating Mobile GIS with enterprise systems uses diverse approaches, each suited to different organizational needs and technical environments.
- **Data extraction, transformation, and loading (ETL):** This approach involves extracting data from GIS, transforming it, and then loading it into other software or systems of the enterprise. Although this method can create duplication of data and may not always be real time, it ensures that data is compatible with the target system.
- **URL-based integration:** This method involves linking systems using URLs to embed GIS maps or functions within enterprise applications. It offers a straightforward approach for adding spatial context to business data without complex programming. Embedding surveys in Experience Builder and Dashboards in the previous chapter are examples of this category.
- **Integration through web services:** Web services enable various applications to communicate with each other over the web using standardized protocols. Using web services, Mobile GIS data can be made directly accessible to other systems. Webhooks and various APIs have been developed based on web services to facilitate integration.
  - **Integration using webhooks:** Webhooks are named for their ability to "hook" into web services, offering a no-code or low-code solution for integrating and automating complex workflows across diverse systems in the cloud.
  - **Custom integration using Python and Jupyter Notebooks:** Python scripts or Jupyter Notebooks can automate the processing, analysis, and movement of GIS data between mobile and enterprise systems. Python, with the ArcGIS API and other extensive libraries, allows tailored integration solutions that can handle complex data workflows or specific analytical needs.

Webhooks

A webhook (also called a web callback or HTTP push API) is a way for an app to provide other applications with real-time information. A webhook delivers data to other applications through HTTPS request (POST) as it happens, meaning you get data immediately—unlike typical APIs where you would need to poll for data frequently to get it in real time. The term was coined by Jeff Lindsay in 2007. In recent years, it has become an increasingly popular method for integration and automation. Webhooks can be used to create automated and integrative workflows, adding new extensibility to Mobile GIS. For example, here are two use cases:
- As the facility manager of a university, you would like to receive email notifications or Teams messages when staff and students report related issues using Mobile GIS.
- As a citizen, you would like to receive a report or permit after the city technician has inspected your property for a project permit you have applied for.

Webhooks (figure 8.1) have the following key terms.
- **Trigger event:** The event that triggers a webhook

- **Payload:** The trigger event data sent to the webhook
- **Payload URL:** The location where the payload will be sent
- **Action:** Something that the webhook does or further invokes

ArcGIS Online and ArcGIS Enterprise support comprehensive trigger events, including when surveys are received and when a hosted feature layer has features added, updated, and deleted. Originally, webhooks had to be configured manually by sending HTTPS requests and parsing the responses. To make setting up webhooks more user-friendly and less technical, ArcGIS software and many vendors provide connectors.

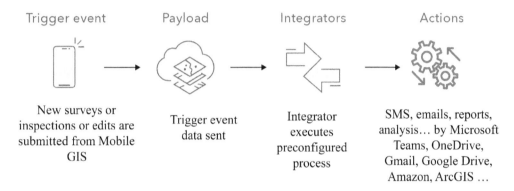

Figure 8.1. Webhooks involve the trigger event, payload, integrators, and actions.

Microsoft Power Automate

Microsoft Power Automate, formerly known as Microsoft Flow, is a powerful cloud-based service for automating workflows between applications and services. It's included in Microsoft Office 365 Business Basic plans and higher. It enables users to create flows, which can connect a wide range of applications, including Microsoft services, such as SharePoint, Outlook, and Teams, as well as third-party apps, such as Twitter, Dropbox, and Google services. Flows can run on a schedule or when triggered. This allows you to automate tasks, such as file synchronization, notifications, data analysis, and more—all without complex coding.

To streamline the use of Power Automate with ArcGIS, ArcGIS provides connectors with prebuilt events and actions. these connectors simplify your workflow by automatically creating the required webhooks within ArcGIS and handling data exchange (figure 8.2). they can receive data payloads and make that dynamic content available for use in subsequent actions within your flow. The commonly used ArcGIS events include the following:
- When a survey response is submitted (triggered by Survey123)
- When a record is created, updated, or deleted (triggered by any mobile, web, or desktop GIS client)
- When an attachment is created, updated, or deleted (triggered by any mobile, web, or desktop GIS client)

The commonly used ArcGIS actions include the following:
- Fetch updates, changes, and deletions from the feature layer
- Create, update, and delete records and attachments
- Find address candidates and geocoding
- Routing
- Create Survey123 reports and get surveys

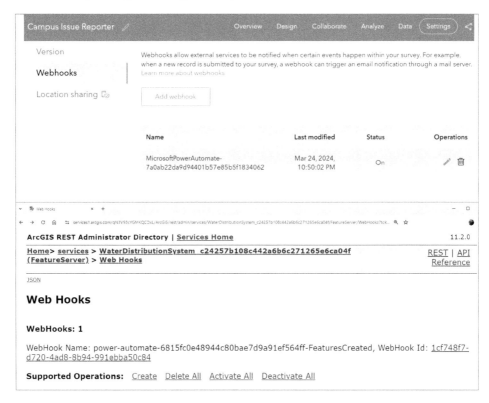

Figure 8.2. *Top*: A Survey123 webhook created by Power Automate with the trigger being when a survey response is submitted. *Bottom*: A hosted feature layer webhook created by Power Automate with the trigger being when a record is created.

Survey123 reports and report templates

As a common workflow pattern, data captured or reviewed by Mobile GIS often needs to be presented in Word or PDF reports or documents. In many instances, it is crucial for these reports, such as official permits, not only to appear professional but also adhere to specific formats to meet legal compliance requirements. Survey123 reports support such workflows by enabling the transformation of your survey responses into richly formatted printable documents (figure 8.3). there are two types of reports in Survey123: individual reports and summary reports. An individual report displays the data for a single survey record, whereas a summary report compiles data from multiple records into one report.

Survey123 reports are generated using report templates associated with surveys. these templates are Microsoft Word files (.docx) containing placeholder text defined in specific syntax. When a report is printed, this placeholder text is replaced with data from the corresponding fields in the survey response. Because these templates are Word documents, they support a wide range of Word's formatting options. these include adding images or logos, text formatting, tables, headers, and footers. The customization flexibility not only helps in creating visually appealing reports but also allows adherence to specific reporting formats required by entities such as government agencies. This flexibility ensures that the reports meet both aesthetic standards and regulatory requirements.

In Survey123 report templates, values of questions can be displayed using expressions such as `${question_name}` or `${floweringtrees / totaltrees}`. these expressions can include methods and parameters. For example:
- `${geopoint1 | getValue:"x"}` has the getValue method to return the x-coordinate of a geopoint.
- `${image1 | size:600:0}` has a parameter to set the width of an image attachment while keeping its aspect ratio.
- `${location | map:"7e2b9be8a9c94e45b7f87857d8d168d6" | mapScale:100000}` has parameters to set the web map item ID and map scale.

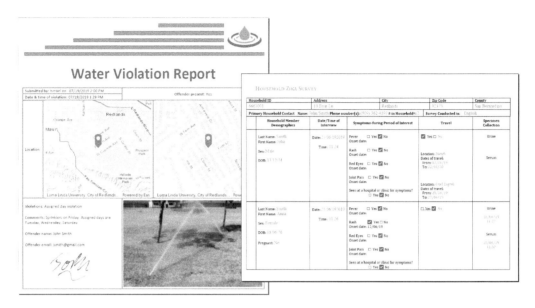

Figure 8.3. Sample Survey123 feature reports.

Tutorial 8: Automating emails and reports and integrating with Microsoft Teams and Microsoft OneDrive

In this tutorial, you will configure a flow to automate Survey123 email notifications, design a feature report template, configure flows to automate Survey123 report generation, save the reports in OneDrive, and post Field Maps messages to Teams.

Data: You will reuse the Hydrant Inspection web map you created in chapter 2 and the Campus Issue Reporter survey you created in chapter 3.

System requirements:
- You must own the Campus Issue Reporter survey to create webhooks and reports on the survey (sections 8.1–8.3).
- You must own the Fire Hydrants feature layer to create webhooks on it (section 8.4).
- Your ArcGIS Online user account needs to have the privileges to generate reports (sections 8.2 and 8.3) and create webhooks (section 8.4).
- You must have a Power Automate account. You can create a trial account if you don't have one.

- You must have a user account for Office 365, which includes Word, Teams, and OneDrive (sections 8.2–8.4).

8.1: Automate Survey123 email notification using webhooks

This section will guide you through configuring a flow to immediately send a recipient email when an issue is reported by the Campus Issue Reporter survey.

1. Open a web browser, go to Power Automate (powerautomate.microsoft.com), and sign in.

2. On the navigation bar on the left, click Create.

3. Choose Automated Cloud Flow from the Start from Blank options.

    The Build an Automated Cloud Flow dialog box appears.

4. Name your flow Survey123 Recipient Email.

    Next, you'll choose a trigger.

5. Type Survey123 in the search bar. Choose When a Survey Response Is Submitted. Click Create.

    Copilot appears to prompt you to sign in to Survey123.

6. If there is already a connection to your ArcGIS Online account, use the existing connection. Otherwise, click Sign In and sign in with your ArcGIS Online account.

7. In the When a Survey Response Is Submitted pane, click Survey, choose the Issue Reporter Survey you created in chapter 3, and leave the other parameters as the defaults.

    If you can't find this survey, you may have multiple ArcGIS connections, and you can click Change Connection and select the correct one.

Next, you will add an action to send out emails.

8. Click the New Step button (plus sign) and click Add an Action.

   Many mail services have connectors that can be used in this flow. For this tutorial, you will use Office 365 Outlook.

9. Search for email and choose Send an Email (V2) in the Office 365 connector.

10. Use an existing connection to Office 365 or follow the prompts to sign in to Office 365.

    Next, you will configure the email recipients, subject, and body.

11. Check Advanced Mode, click the To input box, and click the Insert Token button (lightning bolt), which allows you to enter the data from your survey.

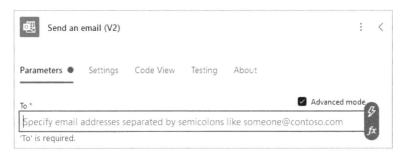

12. Search for email and choose Feature Attributes Your Email.

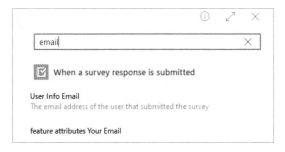

13. For Subject, type Your reported issue has been received.

14. For Body, copy the following text from https://arcg.is/1GLejb1.
    Dear,
    The issue you reported is currently under review. To check its latest status, please click the following link:

    https://survey123.arcgis.com/share/?globalId=&mode=view

    Thank you!
    Campus Management Team

    Next, you will insert the submitter's name, the form item ID, and the GlobalID of the response using dynamic content.

15. In the email body, click the position after "Dear" and add a space. Click Insert Token. In the list of dynamic content, choose Feature Attributes Your Name.

16. Click the position before "?" in the link. Don't add a space. Click Insert Token . In the list of dynamic content, choose Survey Info Form Item Id.

17. Click the position before "&" in the link. Don't add a space. Click Insert Token. From the dynamic content, choose Feature Result globalId.

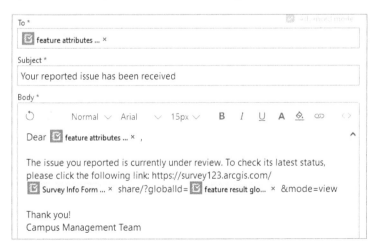

18. Optionally, if you want to be copied on such emails, click Show All Advanced Parameters. In the CC or BCC field, type your email address.

19. Click Save.

Your configuration is complete and is active by default. The webhook is added to your survey automatically.

20. In a new browser tab, go to survey123.arcgis.com and sign in.

21. On your My Surveys page, click the survey you used in the webhook above.

22. On the menu, click the Settings tab and then click the Webhooks section on the left.

This page lists all webhooks you've set up for this survey. The one you just configured is listed here.

23. Under Operations, click Edit to review the webhook configuration.
    - The payload URL is where the survey information will be sent.
    - The trigger event is checked as New Record Submitted.
    - The event data is in the payload that is sent to your flow. This is why you could see all this dynamic content in Power Automate.

24. Click Cancel.

Next, you will test the webhook.

25. Click the Collaborate tab. Next to the link, click Open the Survey in a New Tab.

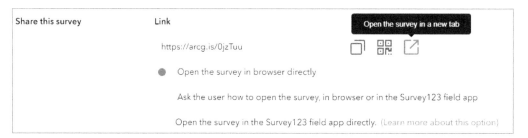

26. Open the survey in a web browser. Complete the form and submit it.

27. Review the email you received.

8.2: Design Survey123 feature reports

In this section, you will create a template for a feature report and use it to produce Word or PDF reports from the Campus Issue Reporter survey you developed in chapter 3. If you do not have the survey or if the survey doesn't have any data, please refer to the tutorial in chapter 3 to create the survey and submit at least one record.

1. In a new browser tab, go to survey123.arcgis.com and sign in.

2. On your My Surveys page, click the Campus Issue Reporter survey you created in chapter 3.

3. On the navigation bar, click Data.

4. On the ribbon below it, click Report.

If you don't have Report on the navigation bar, you may not have the privilege to generate reports. Please contact your ArcGIS administrator or instructor for assistance.

Next, you will create a sample feature report template.

5. Click Manage Templates.

6. In the Manage Templates window, keep Individual Record checked and click Create Sample Template.

This option generates a sample report template that includes the labels and placeholders for all questions in your survey.

7. Click Download Template to download the generated sample template.

8. Find the downloaded template and open it using Microsoft Word.

9. Review the template to understand the basic syntaxes for questions, the map, and the photo attachments.

Next, you will generate a report using the default template.

10. Close the Manage Templates window.

11. Click a record in the data table to select the record.

12. Optionally, change the format to PDF.

13. Click the Preview Sample Report link or the Generate button.

    The preview link will generate a report with a watermark "Sample Report," but it does not cost you ArcGIS Online credits.

14. Find the report generated in your Downloads folder. Open it and review it.

    The report includes a map and one or multiple photos.

    Next, you will enhance the default template by adding a header and a footer.

15. Open the default template that you downloaded in Word.

16. Add a header. In the header, add a logo of your organization.

17. Add a footer. In the footer, add the page number.

    Next, you will open the quick reference page, which provides examples of syntax pertaining to the questions in your survey.

18. On the Survey123 page, click Manage Templates.

19. In the Manage Templates window, click Quick Reference.

    Next, you will improve the report template by zooming the map to a closer scale, increasing the size of the map, and enlarging the photo size.

20. On the Quick Reference page, click the Issue Location question and review the syntax to define map size and scale.

21. In the template .docx file, under Issue Location, change `${issue_location}` to `${issue_location | mapScale: 1000 | size:600:400}`.

22. On the Quick Reference page, click the Photo question and review the syntax to define photo size.

23. In the template .docx file, under Photo, change `${$file | size:460:0}` to `${$file | size:600:0}`.

    Next, you will display the issue type as check boxes.

24. On the Quick Reference page, find the Issue Type question and review the additional syntax for checked check boxes. Copy the following expressions to your clipboard:
    ```
    ${issue_type | checked:"AC"} AC
    ${issue_type | checked:"Lighting"} Lighting
    ${issue_type | checked:"Graffiti"} Graffiti
    ${issue_type | checked:"Pothole"} Pothole
    ${issue_type | checked:"Trash"} Trash
    ${issue_type | checked:"Recyclables"} Recyclables
    ```

25. In the template .docx file, delete `${issue_type}` and paste the expressions you copied.

26. Optionally, insert a table and transfer the Issue Type and Level of Urgency questions into it, placing the former in the left cell and the latter in the right cell.

Issue Type	Level of Urgency
${issue_type	checked:"AC"} AC	${level_of_urgency}
${issue_type	checked:"Lighting"} Lighting	
${issue_type	checked:"Graffiti"} Graffiti	
${issue_type	checked:"Pothole"} Pothole	
${issue_type	checked:"Trash"} Trash	
${issue_type	checked:"Recyclables"} Recyclables	

Description: ${description | appearance:"multiline"}

27. Optionally, for campus name, your name, your email, and report time, place each label and the corresponding expression on the same line.

    ```
    Your Name: ${your_name}
    Your Email: ${your_email}
    Report Time: ${report_time}
    ```

    Next, you will upload the template you enhanced.

28. Save the template as Enhanced_Template.docx.

29. Close the Quick Reference page and return to the Manage Templates window.

30. Click New Template.

31. Select Enhanced_Template.docx to upload it.

32. Click Check Syntax. If there are any errors, correct them in Word, save the template, upload it again, and click Check Syntax Again.

33. Click Save.

    Next, you will use the new template to generate a report.

34. In the Report pane, under Select a Template, make sure Enhanced_Template.docx is selected.

35. Click a record in the data table to select it.

36. Click the Preview Sample Report link or the Generate button to generate a report.

37. Locate and open the newly generated report. Review the report and compare the improvements made to the default report.

8.3: Integrate Survey123 reports with OneDrive

In this section, you will configure a flow to automatically generate a report and save it to OneDrive whenever a new survey response is received. You will use the survey and the report template from the previous section.

1. Return to Power Automate (powerautomate.microsoft.com) and sign in.

2. On the navigation bar, click Create.

3. Choose Automated Cloud Flow under the Start from Blank options.

4. Name your flow Campus Issue Reports to OneDrive.

   Next, you'll choose a trigger.

5. Type Survey123 in the search bar. Choose When a Survey Response Is Submitted and click Create.

6. If there is not a connection to your ArcGIS Online account, click the Sign In as Copilot prompts and sign in with your ArcGIS Online account.

7. From the list of surveys, choose the Campus Issue Reporter survey you used in the previous section. If you have multiple ArcGIS Online connections, you may click Change Connection to select the correct one.

   You can set up a control to generate reports only when a certain condition is met. However, to simplify this tutorial, you will generate reports for all records.

8. Click New Step and click Add an Action.

9. In the Add an Action pane, search for **Survey123** and choose Create Report in the Survey123 connector.

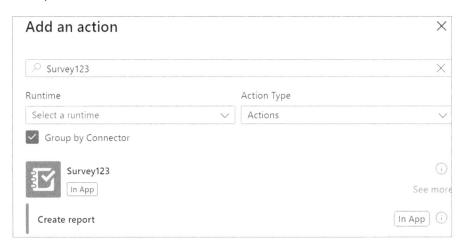

Next, you will configure the parameters for the action. The report will be generated for the record, as identified by the object ID of the submitted record. You will set the report name to follow the format: Issue_objectID_current_datetime.

10. Click Survey and choose the Campus Issue Reporter survey you worked with in the previous section. If you have multiple ArcGIS Online collections, you may need to click Change Connection to select the correct connection.

11. Click Feature Layer and choose Survey, which is the feature layer associated with the survey.

12. For Report Template, choose Enhanced_Template.docx, which is the template you designed in the previous section.

13. Under Feature Object ID, click Insert Token, search for objectid, and choose the feature result objectid.

14. Click Report Name.

15. Type Issue_. Click Insert Token and choose feature result objectid. Type _ and click the fx icon. Under the Function tab, search for current and choose utcNow(). Click Add.

16. For Format, choose pdf.

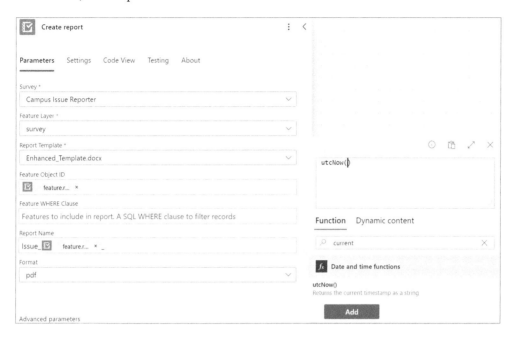

Next, you will add an action to upload the PDF report to OneDrive.

17. Click the New Step button below the Action box and click Add an Action.

18. In the Add an Action pane, search for upload and choose Upload File from URL in the OneDrive for Business connector.

19. In the Upload File from URL pane, click Source URL. Click Insert Token, search for url, and click body/resultInfo/resultFile/url. This is the URL of the first result file.

20. For Destination File Path, specify a OneDrive folder name followed by a slash (/). Click Insert Dynamic Content, search for name, and click body/resultInfo/resultFile/name. Make sure there is a slash between the folder name and the dynamic content.

> **Note:** If you don't have a OneDrive folder or don't know which folder to use, go to OneDrive to create a folder by starting File Explorer, right-click OneDrive, choose New, choose Folder, and specify a folder name.

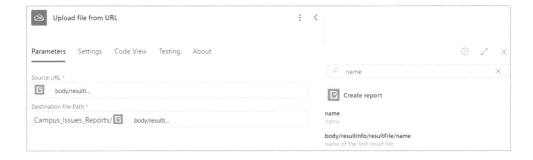

21. Click Save.

    Your configuration is complete. The webhook is added to your survey. Next, you will test the flow.

22. In a web browser, go to survey123.arcgis.com or ArcGIS Online, and sign in.

23. Find the Campus Issue Reporter survey and open it in a web browser.

24. Complete the form and submit it.

25. In File Explorer, go to your OneDrive and find the destination folder you configured. Open and review the PDF report.

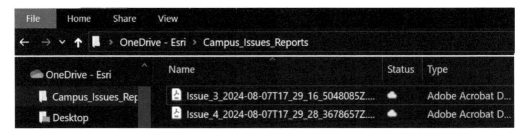

Generating the report can take 20 seconds or longer when it involves many large attachments. If you don't see the report immediately, wait a bit longer.

8.4: Integrate Field Maps with Teams

In this section, you will configure a flow to post a message in Teams whenever a new hydrant inspection is submitted. You will use the feature layer that you created for the Hydrant Inspection Field Maps project in chapter 2.

1. Go to ArcGIS Online and sign in.

2. Click the Content tab.

3. Find the Fire Hydrants layer you created in section 2.1 of chapter 2 and open its item page.

    You must be the owner of the layer to create a webhook for it unless you are the administrator.

4. Click the Settings tab and apply the following settings:
    a. Under Editing, confirm that Enable Editing, Keep Track of Changes to the Data, and Enable Sync are checked.
    b. For What Kind of Editing Is Allowed, confirm that Add is checked.
    c. Click Save.

    Next, you will create a flow in Power Automate.

5. In a web browser, go to Power Automate (powerautomate.microsoft.com) and sign in.

6. On the navigation bar, click Create.

7. From the Start from Blank options, click Automated Cloud Flow.

8. Name your flow Hydrant Inspection Messages to Teams.

    Next, you'll choose a trigger.

9. Type ArcGIS in the search bar. Choose When a Record Is Created in a Feature Layer and click Create.

10. If there is not a connection to your ArcGIS Online account, click the Sign In as Copilot prompts and sign in with your ArcGIS Online account.

11. Click the trigger to open the When a Record Is Created in a Feature Layer pane.

12. Expand the Feature Layer list and click Fire Hydrants/Inspections, which is the feature table you created in chapter 2, section 2.1.

    If you can't find the feature layer, check whether the settings outlined in step 4 have been configured for your Fire Hydrant layer. If not, you may be using a different ArcGIS Online connection and need to select the correct one by clicking the Change connection link.

13. Click Add and click Add an Action.

14. In the Add an Action pane, search for ArcGIS and click See More in the ArcGIS connector.

15. In the list, find and select Fetch Updates, Changes, or Deletions from Feature Layer.

16. Define the parameters for the action:
    a. Click Feature Layer and choose Fire Hydrants/Inspections. If you have multiple ArcGIS connections, make sure you select the correct connection.
    b. Click Get Changes From, click Insert Token, and click Changes URL.
    c. Click Feature Layer ID, click Insert Token, and click Feature Layer ID.

    Notice the fetch action is wrapped inside a For Each loop because the webhook payload may have multiple records included, depending on how actively edited the feature layer is and what the webhook's interval is set to.

17. On the command bar, click Save to save your work.

    Next, you will set up a condition, which will be if the pressure is less than 10 PSI, a message is sent.

18. Inside the For Each loop and under the Fetch action, click Add and click Add an Action.

19. In the Add an Action pane, click Control and click Condition.

    Alternatively, you can search for Condition and then select it.

20. Click the leftmost text box and click Insert Token. Search for Pressure and choose the Pressure field.

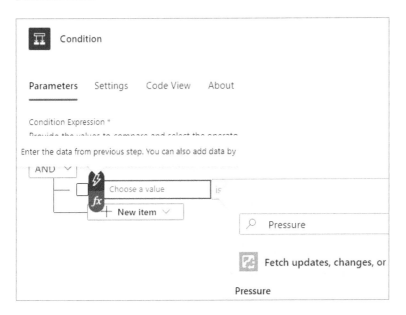

21. For the operator, choose Is Less Than. For the value, type 10.

Next, you will add an action to post a message to Teams if a hydrant's pressure is less than 10 PSI.

22. Under True, click Add and click Add an Action.

23. In the Add an Action pane, search for Post message, and click Post Message in a Chat or Channel in the Teams connector.

24. If you don't yet have a connection to Teams, click Sign In to sign in to your Teams account.

25. In the Post Message pane, apply the following settings:
    a. For Post As, click User.
    b. For Post In, click Channel.
    c. For Team, choose an existing team, or go to Teams, create a new team, and name it Hydrant Maintenance Team.
    d. For Channel, click General or another existing channel.
    e. For Message, type the text and insert the dynamic content of the pressure value and Asset GlobalID as illustrated.

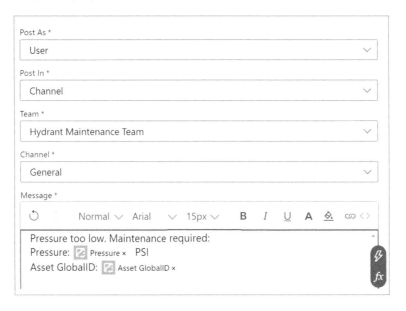

26. Save your flow.

Next, you will test the flow.

27. On your phone or tablet, start Field Maps, sign in, and open the Hydrant Inspection web map you created in chapter 2. If the layer or map has changed, make sure you reload the web map in Field Maps.

28. Tap a hydrant, tap the related button, and tap Add to add a new inspection.

29. For Pressure, specify 5. Optionally fill in other fields.

30. Tap Submit.

31. In Teams, verify the message you received.

In this chapter, you learned how to integrate Mobile GIS with enterprise systems using webhooks. First, you configured a flow to immediately send an acknowledge email to the user who submitted the issue. Second, you designed a Survey123 feature report template and generated a PDF report manually. Third, you configured a flow to automatically generate reports and save them to OneDrive whenever new survey responses are received. Last, you configured a flow to post a service reminder message to a Teams channel whenever Field Maps submits an inspection for any hydrants with low pressure.

The tutorials in this chapter used two triggers: "When a survey response is submitted" and "When a record is created in a feature layer." The former is linked to the survey and, thus, works only with Survey123. The latter is associated with the feature layer and, therefore, can be triggered in various ways, whether the record is created in Field Maps, Survey123, QuickCapture, the Experience Builder Edit widget, or ArcGIS Pro, for example. With the latter, you need to be cautious with the source feature layers and views. If a webhook is set on the source layer, it won't be triggered if the edit is made through its view. And vice versa.

The tutorials in this chapter demonstrated that webhooks facilitate immediate connections with enterprise systems through emails and team messages and showed how webhooks can automate manual tasks, such as report generation and archiving. Integrating Mobile GIS with Microsoft and other enterprise technologies extends Mobile GIS far beyond data collection and visualization, enabling immediate actions, streamlining workflows, and maximizing the value of Mobile GIS.

Assignment 8: Integrate Survey123 with emails and Teams using webhooks

When an issue is submitted through the Campus Issue Reporter survey, the university aims to dispatch the service request efficiently to the relevant departments, using the communication tools each prefers. For example, the Electric Department prefers receiving email notifications for lighting and AC issues, whereas the Campus Cleaning and Recycling Department opts for notifications through messages in a Teams chat group. It is essential to include the type of issue reported in both the emails and Teams messages. Additionally, providing a URL link that allows the recipient to open and review the survey response, including the location and any attachments, is crucial.

What to submit:
- The URL link to your flow. Ensure you share the flow with your instructor as a co-owner.
- Screenshots detailing the condition control, the setup of the email action, and the setup of the Teams message action.
- Screenshots of a notification email received and a Teams message received.

Chapter 9
Virtual reality, augmented reality, and artificial intelligence

Objectives
- Describe virtual reality (VR), web scenes, and scene layers.
- Grasp the concepts and applications of augmented reality (AR) and mixed reality (MR).
- Immerse in and create ArcGIS 360 VR experiences.
- Integrate deep learning models with Survey123.
- Use AI assistants to create surveys using conversational interfaces.
- Create multilingual surveys using automatic translation.

Introduction
This chapter explores the integration of virtual reality (VR), augmented reality (AR), and artificial intelligence (AI) within Mobile GIS, examining how these technologies enhance the way geographic data is perceived, collected, and used. VR immerses users in 3D environments created from GIS data, allowing them to interact within virtual scenarios, whereas AR overlays digital information onto physical environments through devices such as smartphone cameras. Geospatial AI (GeoAI) and deep learning enable automatic identification and categorization of objects within photographs, streamlining data collection and improving the efficiency of spatial inventories. Generative AI (GenAI) transforms user interactions by allowing for natural language processing to simplify tasks such as survey creation.

    The tutorial includes three use cases: The VR use case allows you to experience immersive environments using ArcGIS 360 VR, followed by using Mobile GIS to collect your own data, and finally creating your own VR experiences. The deep learning use case introduces object detection capabilities through a provided survey and data catalog through an Experience Builder web app and guides you in creating your own survey that integrates a pretrained deep learning model. Lastly, the GenAI use case demonstrates how to create and refine survey forms using natural language processing, including multilingual capabilities, to enhance accessibility and user interaction.

Virtual reality

VR is a computer technology that employs headsets to create realistic 3D views, sounds, and other sensations that simulate a user's physical presence in a virtual or imaginary environment. Users equipped with VR gear can "look" and "move" around the artificial world, interacting with virtual features or items.

VR offers a novel way to visualize and interact with GIS data, representing a significant evolution from 2D to 3D GIS maps. Although 2D and 3D maps keep users "outside" the map, VR allows users to step "inside" the map. When used with a headset or helmet, VR immerses users in scenes generated from GIS data. Users experience 3D scenes around them as they walk, turn, and look around. The VR environment is often live and interactive, enabling users to engage with features in the view. This immersive feeling and interactivity bring GIS data closer to the user, enhancing understanding and providing richer insights.

360 VR is a web app (figure 9.1) that enables viewing of 360 VR experiences (3VRs) on desktop PCs, mobile devices, and VR headsets. It is currently browser-based. Users can access 360vr.arcgis.com or log in to their own ArcGIS Online organization or Portal for ArcGIS to view available 3VR items. these 3VRs can depict design alternatives for city blocks for comparison or fictional scenes of historical or futuristic cities. The panoramic images, exported from 3D scenes using ArcGIS CityEngine® or ArcGIS Scene Viewer, are consumable in desktop browsers (where users look around using a mouse), mobile browsers (where users navigate with gyro and touch), and VR headsets (where users look around by turning their heads).

Figure 9.1. ArcGIS 360 VR in a desktop web browser at https://360vr.arcgis.com (*top*), in a mobile web browser (*bottom right*), and in a VR headset with immersive experience (*bottom left*).

Web scenes, scene layers, and 360 VR experiences in ArcGIS

In ArcGIS technology, 3D web maps are called web scenes. In the same way that web maps can contain many layers, a web scene can also host multiple layers, both 2D and 3D.

- **2D layers:** these include feature layers, map image layers, image layers, raster tile layers, and vector tile layers. Most 2D layers will drape over the surface. Feature layers can be enhanced with 3D symbols, have their elevation set as constant values, or use z-values when available.
- **3D layers (scene layers):** these are cached web layers specifically optimized for displaying large quantities of 3D content. Types of scene layers include 3D object scene layers, building scene layers, integrated mesh scene layers, point cloud scene layers, point scene layers, and voxel scene layers.

Most advanced scene layers and web scenes are developed using tools such as CityEngine, ArcGIS Pro, ArcGIS Drone2Map®, and ArcGIS Reality. ArcGIS Scene Viewer facilitates the creation of web scenes with the following general steps:

1. Select a global or local scene.
2. Choose a basemap.
3. Add both 2D and scene layers.
4. Configure layers with styles, labels, and pop-ups. Feature layers can be displayed in 3D, using attributes to control extrusion, size, color, and type.
5. Capture slides to save states such as viewpoint, perspective, daylight, weather, and visibility settings. Slides allow users to quickly navigate to preset states.
6. Save and share the scene.
7. View the 360 VR experience.

Although advanced VR experiences are typically crafted in CityEngine, ArcGIS Scene Viewer also supports the creation of 360 VR experiences. This process transforms a scene into a 360 VR experience and publishes it in ArcGIS Online, converting each Scene Viewer slide into a viewpoint within the VR experience.

Augmented reality, extended reality, and mixed reality

AR is an enhanced version of reality created by superimposing computer-generated information onto the live view of a device (figure 9.2), such as a smartphone or tablet camera. AR is closely associated with Mobile GIS, as a mobile device can retrieve location-based information, including the direction you are facing, the tilt angle of your camera, and the live view in your camera, and overlay this information onto your camera view. Although AR and VR share some similarities, there is a significant difference: AR enhances your current perception of reality, whereas VR replaces the real world with a simulated one.

Figure 9.2. Examples of AR apps developed using ArcGIS Maps SDKs include apps that superimpose underground pipelines (*left*) and overhead electricity power lines (*middle*) onto real-world views. Apps can adjust the heading and elevation settings to align GIS data precisely with the live camera views (*right*).

Extended reality (XR) encompasses all real-and-virtual combined environments and human-machine interactions generated by computer technology and wearables (figure 9.3). The X in XR represents a variable for any current or future spatial computing technologies. XR includes AR, VR, and MR. MR merges real and virtual worlds to produce new environments and visualizations where physical and digital objects coexist and interact in real time. In AR, virtual elements are overlaid in the physical world. In contrast, VR immerses users in a fully virtual environment, disconnected from the physical world. MR bridges these approaches, creating a hybrid experience where augmented and virtual realities converge.

Figure 9.3. Relationships of AR, MR, and VR.

Deep learning and smart assistants in ArcGIS

Broadly speaking, AI enables computers to perform tasks that typically require human intelligence. Machine learning, a key driver of AI, uses data-driven algorithms to learn from data and provide insights. Deep learning is a subset of machine learning that uses several layers of algorithms in the form of neural networks. Input data is analyzed through different layers of the network, with each layer defining specific features and patterns in the data. Deep learning has significantly impacted computer vision, excelling at extracting objects from diverse imagery sources, such as remote sensing data and smartphone photos. This technology has gained substantial attention for its potential in information extraction from imagery.

Geospatial artificial intelligence, or GeoAI, enhances GIS outcomes by using AI subfields, such as deep learning techniques. GeoAI tools excel at identifying meaningful geospatial features from diverse data sources, including text documents and images, streamlining the extraction and analysis process. In ArcGIS, GeoAI can perform many image analysis tasks, including the following.

- **Pixel classification:** Each image pixel is classified into a category, commonly used in land cover classification.
- **Object detection:** This task involves identifying and locating objects within an image by encircling them with bounding boxes (figure 9.4), such as detecting palm trees.
- **Instance segmentation:** This more precise detection method delineates the boundary of each object, useful for tasks such as extracting building footprints.
- **Image classification:** This involves labeling an entire image or specific features within an image, such as categorizing a satellite image as "cloudy" or "clear" or labeling buildings as "damaged" or "undamaged."

ArcGIS enables users to train custom deep learning models and offers ready-to-use pretrained models. Mobile GIS integrates these deep learning models to enhance field data collection workflows, effectively transforming the mobile device camera into a powerful tool that recognizes objects relevant to the task at hand, thereby making data collection more efficient.

For example, Survey123 incorporates smart assistants to augment image-related survey questions, using deep learning packages in the Survey123 field app to enable advanced capabilities, including the following.
- **Smart attributes:** Perform image classification or object detection and provide a real-time preview of attributes during image capture. Once captured, these attributes are stored in the image's EXIF metadata, from which they can be extracted and used to populate other survey questions.
- **Smart annotation:** Use object detection to create annotation graphics on an image, which users can then edit using annotation tools.
- **Smart redaction:** Employ object detection to identify and outline target objects with bounding boxes, and then apply effects to obscure these areas. This is particularly useful for protecting privacy by redacting faces or personal identifiable information.

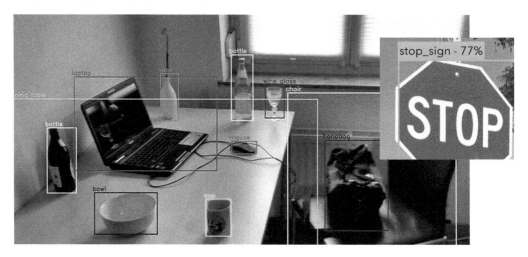

Figure 9.4. Survey123 can integrate deep learning models to automate survey form completion through common object detection.

Generative AI and AI assistants in ArcGIS

Generative AI, or GenAI, is an AI technology that employs generative models to produce text, images, videos, or other data types, typically in response to user prompts. these models assimilate the patterns and structures from their input data to create new, similar outputs. Large language models (LLMs), a specialized subset of GenAI focused on text generation, are trained on extensive datasets and can generate outputs ranging from emails to comprehensive reports. This innovation has transformed AI into a more interactive and accessible tool, revolutionizing how tasks are approached by offering creative reasoning through natural language interfaces. Since their debut in late 2022, GenAI applications have driven companies across various industries to integrate this technology into their business processes, significantly boosting productivity and efficiency.

In the realm of Mobile GIS, advances have particularly focused on automating and enhancing workflows through human-computer interaction. For example, AI assistants in ArcGIS, such as Survey123 smart assistants (figure 9.5), streamline the survey design process by enabling users to create surveys conversationally, similar to interactions with platforms such as ChatGPT. This capability allows users to quickly draft surveys without navigating complex interfaces. Additionally, the integration of automatic translation in Survey123 uses the latest LLM advancements to translate survey content into multiple languages effortlessly, enhancing global accessibility and usability. Although generative AI is not without its flaws, the integration of GenAI within Mobile GIS is empowering a wide array of users—from public-sector workers to community leaders—to use Mobile GIS more effectively.

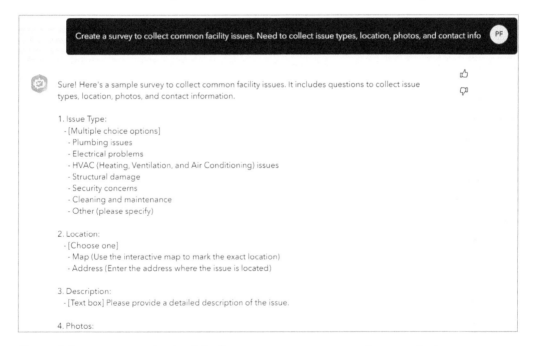

Figure 9.5. AI assistants for ArcGIS allow users to create survey forms and other content types through natural language conversations, simplifying the design process.

Tutorial 9: Create VR experiences, use deep learning packages, and explore generative AI in Mobile GIS

This tutorial explores the integration of VR, deep learning, and GenAI within Mobile GIS applications, structured around three use cases, each focusing on different technologies and their applications.
- **Use case 1** (sections 9.1–9.3): Experience 360 VR. Collect field data using Mobile GIS and create a VR experience using Scene Viewer.
- **Use case 2** (sections 9.4 and 9.5): Develop a survey to perform a photo-based resource inventory and automatically detect common objects in the photos.
- **Use case 3** (sections 9.6 and 9.7): Use GenAI assistants with Survey123 to create surveys using natural language and to make the forms multilingual.

System requirements:
- Section 9.5 requires a Windows computer to run Survey123 Connect.
- Section 9.6 requires an ArcGIS Online Administrator role.

9.1: Experience VR using browsers, smartphones, and headsets

1. On your mobile device, scan the following QR code.

   A 360 VR experience will be downloaded and the first viewpoints displayed.

2. If prompted by a Panorama View symbol, tap it.

3. If prompted that ArcGIS Online would like to access the motion and orientation of your device, tap Allow.

4. Rotate the device around and up and down to change orientation and tilt angles.

Chapter 9: Virtual reality, augmented reality, and artificial intelligence   213

5. Tap a different viewpoint—for example, Parking Lot or Birds View.

    The scene changes perspective. The operational layers, basemap layer, and weather change, too.

6. Tap a feature, such as a tree, to view a pop-up.

    The features in the VR are more than just images; they come equipped with attributes.

    Next, you will explore the 360 VR Experiences gallery.

7. Tap the Content Browser button (nine dots) located in the upper left of the page or visit https://360vr.arcgis.com.

    The 360 VR Experiences gallery will appear, showcasing the available VR experiences that you can access.

8. Explore additional VR experiences, such as Pasadena Redevelopment and City of Boston.

These experiences may include scenarios, such as current scene and future design in the Pasadena Redevelopment VR experience. If multiple scenarios are available, click on them to explore them and understand the possible applications of each scenario's capability.

Next, you will explore VRs using your desktop browser.

9. On your desktop browser, navigate to https://arcg.is/rajef and explore the Fun Park VR.

10. Click and drag to navigate through the scenes both horizontally and vertically. Click on the viewpoints to switch between slides.

11. Tap the Content Browser button in the upper left of the page or visit https://360vr.arcgis.com.

12. Explore additional VR experiences, such as the ones you just explored, using your device.

Next, if you have a headset, you will experience the VRs using one. This tutorial uses the Meta Quest 2, but other headsets may have different features or controls.

13. With the Meta Quest headset, press the Home button on your controller to open the home menu.

14. Start Meta Quest Browser and navigate to https://arcg.is/rajef.

15. If prompted with the message Allow ArcGIS Online to Open an Immersive Experience and Access Hand Tracking, click Allow.

The VR experience downloads and displays.

Chapter 9: Virtual reality, augmented reality, and artificial intelligence **215**

16. Move your head side to side and up and down to fully immerse yourself in the VR experience.

17. Choose various viewpoints to explore different scenes within the VR experience.

18. Click the Close (X) button located beneath the Viewpoints bar to exit the Fun Park VR experience.

    Next, you will explore the 360 VR gallery.

19. In the Meta Quest browser, navigate to https://360vr.arcgis.com.

20. Explore additional VR experiences, such as Pasadena Redevelopment and City of Boston.

9.2: Collect data for your VR experience

This section provides guidance on collecting data for the Fun Points layer within the Fun Park VR experience you explored in section 9.1. The data you collect will be used to create your own VR experience in the next section.

1. You may collect data using ArcGIS Field Maps or ArcGIS QuickCapture or both by scanning the QR codes, for Field Maps (*top*) and QuickCapture (*bottom*).

The web map for Field Maps and the QuickCapture project both use the same Fun Points layer, ensuring that data collected using either Field Maps or QuickCapture is stored in the same layer.

2. Visit a nearby park or community to collect 30 or more points. For each point, select the feature type and specify the size and orientation.

> Note: The layer is configured to allow visibility of all data to everyone, facilitating team collaboration. For instance, different members of the class can be responsible for various parts of the park or different types of features.

To expedite the tutorial, the size and orientation values need not be precise. For an optimal VR experience, it's best to collect points that are close together rather than spread out.

9.3: Author a web scene and a VR experience

In this section, you will use the data collected in the previous section to create a web scene and develop a VR experience.

1. In a web browser, navigate to ArcGIS Online, and sign in.

2. On the upper ribbon, click Scene.

3. Click New Scene.

4. On the map toolbar, click Basemap, choose Imagery Hybrid, and then close the Basemap gallery.

   Next, you will add the Fun Point layer to the scene.

5. On the Designer toolbar (dark), click Add Layers and choose Browse Layers.

6. Click My Content and click ArcGIS Online from the list. Search for fun points sample owner:GTKMobileGIS.

7. Click Add and click Done.

8. Zoom to the area where you collected data and tilt the scene to a perspective you like.

   To navigate the scene, press and hold the left mouse button to pan, and press and hold the right mouse button to rotate.

   The layer is configured so that everyone can see all users' data. If the data in your area becomes too crowded, you have the option to exclude other users' data.

9. Optionally, if you want to exclude data collected by other users, apply the following settings:
    a. On the Designer toolbar, click the Layer Manager button.
    b. Hover over the Fun Points layer and click the Options button (three dots).
    c. Choose Layer Properties.
    d. Under Filter, set the Creator field and click your username.
    e. Click Done.

Next, you will style the Fun Points layer using the type, size, and rotation attributes.

10. In the Layer Manager pane, click the Fun_Points layer.

11. For the main attribute to visualize, choose ObjectType. For the drawing style, choose 3D Types and click Options.

12. Under All Markers, for Size, choose ObjectSize with ft as the unit. For Rotation, choose the ObjectRotation field.

Now the points display as 3D symbols of varying sizes.

Next, you will assign specific symbols to each point type.

13. In the Attribute Values list, click the Bush category and click the Marker box.

14. In the Symbol Picker window, click Basic Shapes and click Vegetation from the list. Click Flannelbush or another symbol of your choice and click Done.

The symbol icons are displayed in alphabetical order. You can switch from Grid view to List view to find the symbol icons easier. The bush-type points now are updated in the scene with the new symbol.

15. Repeat the previous two steps to choose symbols for the remaining point types, starting with the Vegetation group and then moving on to others. Feel free to choose different symbols.
    a. For Flower, in the Vegetation group, choose Rhododendron Azaleas.
    b. For Tree 1, in the Vegetation group, choose Flowering Dogwood and click Done. Then click the Color button to choose pink (or use the color code #FFBEBE).
    c. For Tree 2, in the Vegetation group, choose Flowering Dogwood.
    d. For Rock, in the Street Scene group, choose Rock 1.

e. For Round Bench, in the Street Scene group, choose Park Bench 4.
f. For Bench, in the Street Scene group, choose Park Bench 2.
g. For Table, in the Street Scene group, choose Picnic Table.
h. For Boat, in the Transportation group, choose Motorboat.
i. For Car, in the Transportation group, choose Tesla P7.

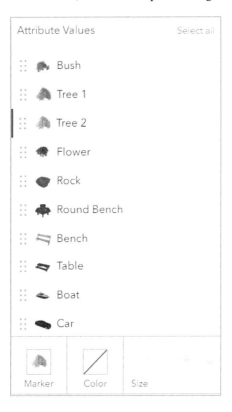

16. Click Done and then click Done again to finalize the styling of the point layer.

17. On the Designer toolbar, click Save. Name your scene (for example, My Fun Park or My Community, followed by your name) and click Save.

    Next, you will learn how to edit the fun points if you need to modify their sizes and orientation values.

18. On the Designer toolbar, click Edit.

19. In the Editor pane, under Edit Features, click Select.

20. Click on an object that you added—a bench, for example—on the map.

The object is highlighted with interactive handles, appearing as circles.

> **Note:** This layer is configured so that editors can edit only features they own. If the interactive handle is not visible, it means you do not own the feature. Please select a feature that you own to make edits.

21. To edit the object, perform the following actions:
    a. Drag the outer circle in or out to resize the bench.
    b. Rotate the outer circle to reorient the bench.
    c. Drag the inner circle to reposition the bench.

    The ObjectRotation and ObjectSize attributes will update as you resize and rotate the bench.

22. Optionally, continue editing additional points or add new points as needed.

    Next, you will create several slides for your VR experience.

23. On the Designer toolbar, click the Slide Manager button to open the Slide Manager pane.

24. Navigate to a scene view you want to feature in the VR experience. You may change the basemap, daylight settings, weather effects, the viewing extent, and the perspective.

25. Click Capture Slide.

    A new slide with a thumbnail of the view is added to the list.

26. Click the slide name to edit it and enter a new name of your choice.

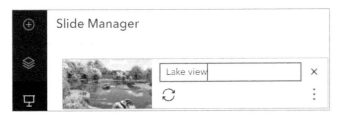

27. Repeat the previous three steps to capture two or more slides.

28. Click Save to save your scene.

    Next, you will export the scene and the slides into a VR experience.

29. On the Designer toolbar, click the Create App button. Click 360 VR Experience.

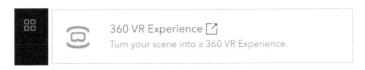

    The Create 360 VR Experience page appears.

30. Optionally, change the name of the VR experience and then click Publish.

    The process may take a few minutes to complete. Each slide in the web scene is converted into a viewpoint in the

31. Once the VR experience is created, click Open in 360 VR.

32. Optionally, explore the VR experience on your phone. You can generate a QR code for easy access.

33. If you have a VR headset, optionally explore the VR experience using it. Consider generating a short URL to simplify access to your VR experience.

9.4: Perform photo-based inventory using an object detection model

In this section, you will use Survey123 to inventory resources by capturing photos and employing an object detection model to recognize and catalog objects, saving this information to a feature layer. An Experience Builder app will then be used for you to search for and locate these objects and their associated photos.

1. On your mobile device, scan the QR code.

   A survey named Photo Inventory will be downloaded, along with an object detection deep learning package, and then the survey form will open.

2. On the survey form, use your current location or select a different location.

3. Under the Take a Photo option, tap the Camera button.

4. Point the camera at common objects, such as keyboards, mice, cups, books, computer monitors, and bottles. Observe how the names of the detected objects display immediately in the camera preview, along with the confidence percentages and bounding boxes.

   For a complete list of objects the model can detect, refer to https://arcg.is/ynf5X0. This model is not recommended for use in production surveys.

5. Capture the photo. Observe the types of detected objects are listed as comma-separated values (CSV) in a text box and as check boxes, along with the count of object classes detected.

6. Under EXIF Image Description, view the JSON file that details the object names and their bounding boxes.

   This JavaScript Object Notation (JSON) is generated by the deep learning package based on the photo you captured. It feeds information to the detected object class names text question and the check box question on the form. Typically, this JSON should be hidden on the form. However, it is displayed here to illustrate the logic behind the smart attributes and how the form questions are formulated. Formulas will be explained in the next section.

   Sometimes, the model may incorrectly identify objects, and the check boxes allow you to make corrections.

7. Uncheck any object class names that are not accurately represented in the photo.

8. Submit the survey.

9. Repeat the process to capture and submit several more photos.

Next, you will review the resources inventoried using an Experience Builder web app.

10. On a desktop or mobile device, navigate to https://arcg.is/1rS5yf0.

11. Under Find Photos Containing, click the drop-down arrow and search for an object, such as chair.

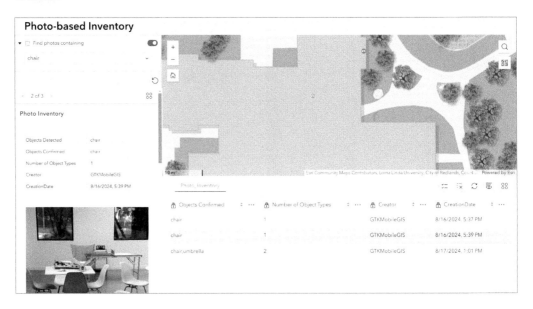

12. Observe how the table updates to show only features containing these objects, the map zooms in and highlights these features, and their pop-ups along with the photos are displayed in the feature Info widget.

The next section will demonstrate how to create a survey that collects photos and automatically recognizes objects within them.

9.5: Create a photo-based inventory survey using a deep learning model

This section will create the survey you used in the previous section.

This section requires a Windows computer to run Survey123 Connect.

1. Start Survey123 Connect and sign in.

2. Download the Excel file from https://arcg.is/0PuiLX1 and drag it into Survey123 Connect.

   The survey will automatically be created and displayed in preview mode.

   Next, you will examine the details of the Excel workbook.

3. On the left toolbar, click XLSForm.

   The Excel workbook opens.

4. In the Excel workbook, review the following questions and their calculation formulas:
   - The geopoint question to collect locations.
   - The image question named smart_img to collect photos.
   - The text question named object_names. It pulls classNames from the smart_attributes_results question. This essentially represents the names of objects detected in the photo.
   - The calculation named num_objects. It counts the number of object classes checked. The value is displayed in the label of the objects_checkboxes question.
   - The question named objects_checkboxes. It displays the detected object names as check boxes. This question has a filter to show only object classes that are detected in the photo.
   - The text question named smart_attributes_results. This is the last question on the sheet and is used to pull the ImageDescription JSON from the smart_img question.

5. Find the smart_img question and observe the bind::esri::parameters column, which reads `smartAttributes=CommonObjectDetection&cameraPreview=true`

   `smartAttributes=CommonObjectDetection` links the image question to an object detection model, allowing it to extract values based on the objects the model detects in the image. The specified model here is CommonObjectDetection.

   `cameraPreview=true` enables a real-time preview where objects detected by the model are highlighted with bounding boxes in the live camera view.

6. Close the XLS form.

7. Click Publish, click Publish Survey, and click OK when publishing is completed.

   Next, you will link the survey to the Common Object Detection deep learning package.

8. On the lower toolbar, click Linked Content and then click Link Content.

9. Click Deep Learning Package.

10. In the Link Deep Learning Package window, click Filters and turn off all the filters.

11. Type Common Object Detection in the search bar.

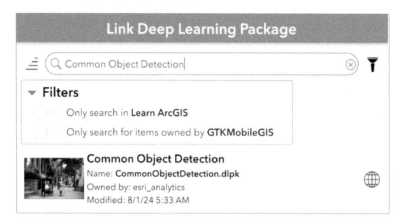

12. Click the Common Object Detection model (CommonObjectDetection.dlpk) and click OK.

13. Click the Download button to download the model to your computer.

    The deep learning package is saved to the survey's media folder.

14. On the left toolbar of Survey123 Connect, click the Files button to open File Explorer.

15. In File Explorer, go to the media folder and verify the CommonObjectDetection model files are there. Close File Explorer.

    Next, you will publish and share the survey.

16. In Survey123 Connect, click Publish and click Publish Survey. Click OK when publishing is completed.

17. On the left toolbar, click the More Actions button (three dots). Then, click Manage in Survey123 Website.

18. Click the Collaborate tab.

    The deep learning model functions only in the Survey123 native app, not in the browser app.

19. For Link to the Survey, choose Open the Survey in the Survey123 Field App Directly and click the QR code.

20. Use your mobile device to scan the QR code and test the survey, as done in the previous section.

21. On a desktop or your mobile device, visit https://arcg.is/1rS5yf0 and search for the objects you just collected, as done in the previous section.

9.6: Enable Survey123 smart assistant and autotranslation

> **Note:** This section requires an ArcGIS Online administrator to enable Survey123 Assistant and autotranslation.

1. Sign in to ArcGIS Online.

2. Click the app launcher (nine dots) and click Survey123.

3. From the menu, click Organization and then click Settings.

4. On the left, click Extensions.

5. Enable the options for Survey123 smart assistant and Auto Translate.

6. Click Save to apply the changes.

9.7: Explore Survey123 smart assistant and autotranslation

This section requires a creator account, not an administrator account.

1. Sign in to ArcGIS Online.

2. Click the app launcher and click Survey123.

3. Click New Survey.

4. Under Blank Survey, click Get Started.

5. In the lower right, click Survey123 Assistant.

   An introduction about Survey123 Assistant appears.

6. Click Continue.

7. In the message area, describe the survey you want to create using natural language. Here are some examples:
   - Create a survey to collect common facility issues. Need to collect issue types, location, photos, and contact info.
   - Create a survey to collect hazardous trees in the city of <your city>, including questions to capture location, photo, tree species, and a quick risk assessment.

   Don't worry about typos; the language model used can typically understand sentences even if they are not perfectly constructed.

8. Review the list of survey questions suggested by the assistant and request specific refinements, such as the following:
   - Question 1 should list 10 common facility issues, including broken glass and light bulb failure.
   - Add a question to ask for issue details.
   - Question 3 should include a list of common California tree species.

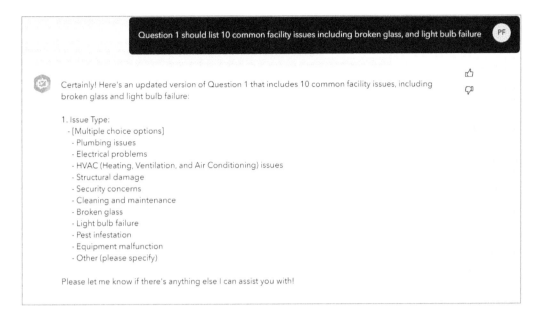

9. Click Generate to create the survey.

10. Set the survey title appropriately—for example, Report Facility Issues or Assess Hazardous Trees.

11. Optionally, refine the survey manually to better meet your specific needs.

    Next, you will add a new language to the survey to enable multilingual support.

12. In the Design pane, click the Options tab. Then click the Manage Survey Languages button.

13. Click Add Language, choose a language you want to add, and click Add.

14. Click Auto Translate and then click Continue.

15. Choose the target languages and click Translate. Allow the translation to occur before moving on.

    Your form is now translated. The source and target languages display side by side, allowing you to manually edit the translation if necessary.

16. Click Done.

    Next, you will preview the form in multiple languages.

17. On the lower-right toolbar, click the Preview tab to see your survey.

18. At the top of the page, click the language selector to switch to a different language.

19. Click Close Preview.

20. Optionally, publish the survey and test the survey.

This tutorial explored three emerging technologies within Mobile GIS: VR, deep learning, and generative AI.
- **VR:** You began by experiencing VR using phones and desktop browsers, followed by an immersive experience with headsets, showcasing VR's engaging potential. Next, you used Field Maps and QuickCapture to collect your own data, applied 3D symbols, and created your own simple VR environment.
- **Deep learning:** You explored object detection capabilities using a provided survey and reviewed the data collected through an Experience Builder web app. You then demystified the process by creating your own survey, integrating a pretrained deep learning package. This capability allows for automatic identification and categorization of objects

within photographs, streamlining data collection and improving the efficiency of spatial inventories.
- **Generative AI:** You first assumed the role of an administrator, enabling the Survey123 smart assistant and its autotranslation features. You then assumed the role of a creator, used the smart assistant to create a form through natural language input, and used autotranslation to make the form multilingual. The AI assistant enhances user interaction by supporting easy form creation and multilingual deployment, making Mobile GIS more accessible to a diverse audience and catering to both novice and expert users.

The VR experience created in this tutorial is straightforward but more advanced scene layers and VR experiences can be crafted using tools such as ArcGIS Pro and CityEngine. The deep learning model used in this tutorial is basic and may occasionally misidentify objects, highlighting the need for improved accuracy. ArcGIS offers tools for training and enhancing deep learning models to address these issues. Additionally, the survey and translation AI assistants generated are not flawless and may require manual refinements. However, these technologies are rapidly evolving and will make Mobile GIS more engaging, easier to use, and more accessible.

Assignment 9: Create a VR experience of a fun community

Create a 360 VR experience representing a fun community, incorporating various geographic elements and interactive features.

Requirements:
- Choose a location, such as a street, residential area, park, or any other location.
- Include at least a point feature of your own. The layer should use 3D symbols that are driven by attributes such as ObjectType, ObjectSize, and ObjectRotation, as demonstrated in the tutorial.
- The ObjectType should have some coded values different from the Fun Points layer from the tutorial. (Hint: You may copy the Fun Points layer and then modify the coded values.)
- Include at least a line layer to represent elements such as roads, pathways, fences, pipes, or boundaries.
- Create at least two slides within your web scene, each showcasing different viewpoints with different basemaps and weather conditions.

What to submit:
- The URL to your VR experience.

Chapter 10
Developing custom Mobile GIS apps

Objectives
- Describe the approaches to developing custom mobile apps.
- Learn the basics of ArcGIS Maps SDKs for Swift, Kotlin, and .NET MAUI.
- Query map layers and handle user interactions using Maps SDKs.
- Develop responsive browser-based apps using JavaScript and .NET MAUI.
- Develop cross-platform native apps using the ArcGIS Maps SDK for .NET MAUI.

Introduction
In earlier chapters, you were introduced to building Mobile GIS solutions using off-the-shelf mobile apps, web apps, and low-code strategies with Arcade and webhooks. Although these methods provide a cost-effective way to meet most common Mobile GIS application requirements, they might not fulfill all the specific needs of your project. In such cases, you will need to create your own custom mobile apps. This chapter will guide you through the process of developing responsive, custom mobile apps using JavaScript, .NET MAUI, and map software development kits, or ArcGIS Maps SDKs.

Mobile app development approaches
Choosing a mobile app development strategy depends on several key factors, including the expertise of the development team, the specific functionality required by the application, the target platforms, and the available budget. Mobile app development encompasses the following approaches, each with its distinct characteristics and implications.
- **Browser-based approach:** This method involves creating apps that operate within web browsers using HTML, JavaScript, and Cascading Style Sheets (CSS). Considering the ubiquity of web browsers on both mobile and desktop platforms, this approach can reach a broad audience, offering a cost-effective and faster development cycle compared with native app development. However, its limitations include restricted support for offline functionality and limited access to a device's native features, potentially leading to a user experience and performance that may not equal that of native apps.

- **Native-based approach:** Native apps are downloaded and installed directly on mobile devices, providing broad access to a device's hardware and resources, which enables a more seamless user experience compared with browser-based apps. Developing native apps requires proficiency in platform-specific programming languages, such as Swift for iOS and Kotlin for Android. Swift, developed by Apple, offers seamless integration with iOS frameworks and APIs, allowing developers to fully work with iOS-specific features such as Core Location and Maps. Kotlin, the preferred language for Android development, supports efficient, reliable mobile apps optimized for Android's diverse ecosystem, ensuring broad compatibility and strong performance across multiple devices. Although native apps often deliver superior performance and user experience, they tend to be more costly and time-consuming to develop, as each platform requires its own tailored solution, increasing the effort required for multiplatform coverage.
- **Hybrid-based approach:** This strategy aims to combine the best aspects of browser-based and native approaches by integrating native components with HTML, JavaScript, and CSS. The resulting applications are technically native but use web technologies for much of their functionality. Although this approach offers numerous benefits, it also presents challenges, such as potentially slower performance compared with native apps because of the reliance on a browserlike component. Additionally, although hybrid apps can access device features, they may not fully exploit specific platform capabilities as native apps can.

Each approach offers distinct advantages and limitations, making the choice of development strategy a crucial decision that should align with the project's functional and user experience requirements, the development team's expertise, and available resources.

ArcGIS Maps SDKs

ArcGIS software provides Maps SDKs for developing mapping and spatial analysis apps across web browsers, native devices, and game engines. these SDKs support browser, native, and hybrid approaches for building custom Mobile GIS apps. The ArcGIS Maps SDK for JavaScript is designed for both browser-based and hybrid-based approaches. The ArcGIS Maps SDKs for Kotlin™ and for Java™ are tailored for native Android apps, whereas the ArcGIS Maps SDK for Swift supports native apps for iOS and iPadOS. Additionally, the ArcGIS Maps SDKs for .NET MAUI, Qt™, and Flutter™ are versatile, capable of building apps for multiple platforms.

Each SDK (figure 10.1) includes an application programming interface (API), API reference, documentation, tutorials, samples, and integration with integrated development environments (IDEs). All these SDKs interact with ArcGIS Online and ArcGIS Enterprise through ArcGIS representational state transfer (REST) APIs, offering similar core functionalities. these functionalities encompass the following:

- Adding mapping, querying, editing, spatial analysis, and other functions
- Delivering high-performance 2D and 3D visualization
- Integrating with ArcGIS Online and Enterprise through REST APIs
- Managing secure workflows
- Supporting offline maps, data, routing, and geocoding
- Providing advanced client-side visual analysis and geometric operations

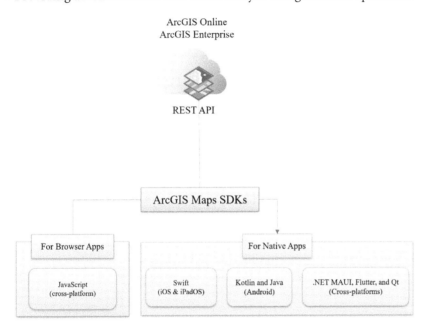

Figure 10.1. The ArcGIS location suite provides the REST API and Maps SDKs for developing browser-based and native apps.

To make this book relevant to both iOS and Android platforms within the limited space, the tutorial section focuses on the cross-platform development option using ArcGIS SDK for .NET MAUI. However, it's important to note that the ArcGIS Maps SDKs for Swift and Kotlin are often the preferred choices for developers looking to fully harness Mobile GIS capabilities on iOS and Android platforms, respectively. The Swift Maps SDK allows iOS developers to integrate ArcGIS's mapping and spatial analytics capabilities directly into their apps, making use of Swift's modern features to deliver an intuitive user experience. The SDK optimizes iOS-specific functionalities, ensuring apps are powerful and efficient. Similarly, the Kotlin Maps SDK provides Android developers with the tools needed to build highly responsive and reliable GIS applications. By using Kotlin, these apps can perform complex spatial analysis with seamless interaction and visualization capabilities, ensuring compatibility and high performance across Android's extensive range of devices. Together, these SDKs empower developers to create sophisticated native Mobile GIS applications that push the boundaries of what mobile devices can achieve in mapping and spatial analysis.

JavaScript, HTML, CSS, and responsive web design

JavaScript, HTML, and CSS are the languages for browser-based app development, integral to nearly every web page in existence today. these technologies work together in the following ways:
- HTML organizes the content.
- CSS defines the style.
- JavaScript introduces interactivity and dynamic features.

Responsive web design ensures that a website looks and functions well on a variety of devices and screen sizes, from smartphones to tablets, and desktop computers. This design methodology aims to provide an optimal viewing experience—easy reading and navigation with minimal resizing, panning, and scrolling—across a wide range of devices. CSS media query is the main technology behind responsive web design, allowing developers to apply different styling rules based on the characteristics of the device viewing the website, such as its width, height, orientation, and resolution. This is fundamental for browser-based Mobile GIS app development.

Here are some typical CSS media query examples:

```
@media (orientation: portrait) {
/* whenever the width is shorter than the height */
}
@media (orientation: landscape) {
/* whenever the width is longer than the height */
}

@media (max-width:768px) {
/* tablets */
}
@media (max-width:414) {
/* Phones */
```

While developing JavaScript apps, you will need an IDE to be more productive. there are many free JavaScript IDEs available, including Microsoft Visual Studio Code (VS Code), WebStorm, and Sublime Text. IDEs can help developers write code more efficiently by offering IntelliSense, which provides automatic code completion, syntax highlighting, and context aware suggestions. Most web browsers offer developer tools that can display JavaScript errors in the console, set breakpoints, monitor network traffic, and inspect HTML elements and styles.

After an app is developed, it needs be deployed to a production environment, which involves transferring the HTML/CSS/JavaScript files from a local development environment to a web server. This makes the app accessible to mobile and web users over the internet. Microsoft Internet Information Services (IIS) is a popular choice for hosting apps on Windows servers. Deployment to IIS typically requires copying the files to the server's designated directory—for example, C:\inetpub\wwwroot—and configuring the site settings within IIS Manager. Beyond IIS, various other web servers and hosting options are available for deploying web apps, including Apache and Nginx on Linux, or cloud-based platforms, such as Microsoft Azure, Amazon Web Services (AWS), and Google Cloud.

Microsoft .NET MAUI

Microsoft .NET MAUI (Multi-platform App UI) is a framework designed for creating cross-platform native mobile and desktop apps. It enables the compilation of a single code base into apps for iOS, Android, macOS, and Windows (figure 10.2). Once your mobile app is developed, you can distribute it through the App Store and Google Play.

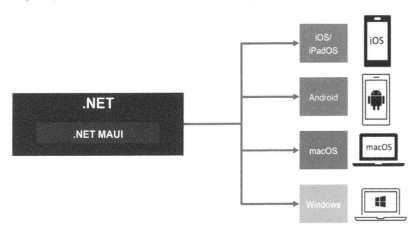

Figure 10.2. Microsoft .NET MAUI is a framework for building cross-platform native apps for iOS, Android, macOS, and Windows from a single code base.

.NET MAUI typically employs extensible application markup language (XAML) to define app content and styles and uses C# to implement logic. To draw a parallel with web development, XAML is analogous to HTML and CSS, whereas C# corresponds to JavaScript. For a simple app created using the ArcGIS Maps SDK .NET MAUI App Template, the app content is organized in MainPage.xaml, the styles are defined in resources\Styles\Styles.xaml, and the logic is implemented in MainPage.xaml.cs.

The IDE for developing .NET MAUI apps is Microsoft Visual Studio. Visual Studio (figure 10.3) provides a comprehensive set of tools and features specifically designed to facilitate the development, debugging, and deployment of .NET MAUI apps across various platforms. Visual Studio is available in Professional, Enterprise, and Community editions, with the latter being free and recommended for use with tutorial sections 10.4–10.6.

Figure 10.3. The Microsoft Visual Studio IDE is equipped with specialized tools for developing, debugging, and deploying .NET MAUI apps across multiple platforms.

When developing mobile apps using Visual Studio, testing and debugging are essential steps to ensure your app functions correctly on every operating system you target. Visual Studio allows you to run tests on your app for both Android and iOS.
- **For Android:** Visual Studio provides the Android Device Manager (ADM), for creating Android Emulators. You can run your app in the emulators and debug your app.
- **For iOS:** Visual Studio facilitates testing through the iOS Simulator on macOS or through connected physical iOS devices or an iOS simulator. Visual Studio in Windows typically needs to pair with a Mac remotely to compile and test iOS apps.

Case studies: Delivering air quality information at your fingertips

Air pollution is a significant public health concern in many places, particularly in Southern California. Understanding when and where air quality is good or bad is crucial for people planning their activities, especially for those who are sensitive to air quality or ill.

To meet these needs in Southern California, the South Coast Air Quality Management District (AQMD) has developed a modern, innovative Mobile GIS app (figure 10.4) that delivers essential air quality information directly to users across several counties. This mobile app is available on the App Store and Google Play. It features an air quality map that displays the current regional conditions and provides detailed air quality information for the current time and the coming days at your current location, home, work, or any other places you have set up. The information includes the air quality index, pollution levels, weather conditions, and temperature. It also identifies the main pollutants affecting the area, further educating the public about environmental concerns. Additionally, the inclusion of alternative fuel station locations supports environmentally friendly transportation choices, aligning with broader sustainability goals.

Figure 10.4. The South Coast Air Quality Management District's Mobile GIS app, developed with ArcGIS Maps SDKs, provides real-time air quality information and forecasts across Southern California, available on the App Store and Google Play.

The air information and web map are hosted in ArcGIS Online. The app was developed using Maps SDKs. It queries the layers for your current location and any other places you have set up and then displays the information as gauges and text with different colors to represent quality levels, enhancing the user experience by making complex information accessible and understandable.

The app exemplifies the use of Mobile GIS in environmental monitoring, public health, and public information services. It enables residents to make informed decisions about outdoor activities based on air quality forecasts.

Tutorial 10: Develop a game using JavaScript and .NET MAUI

In this tutorial, you will develop a game that generates questions about countries around the world, asks users to identify and tap the country on a map, and informs them whether their answers are correct or incorrect. The app needs to be responsive to both landscape and portrait orientations to best use the screen real estate. You will develop the same game twice: once as a browser-based version using JavaScript (sections 10.1–10.3) and again as a cross-platform native app using .NET MAUI (sections 10.4–10.6).

Data: You are provided with a web map (https://arcg.is/0XLyGn). It has a feature layer of world countries with generalized geometries (figure 10.5). You are also provided with a Chapter 10 data folder, which can be downloaded from https://arcg.is/1Orn9a1. Download and extract the zip file to a folder of your choice. It has HTML, XAML, and C# files that will help you design the app.

System requirements:
- Sections 10.1–10.3: A JavaScript IDE is required. If you don't have one, Microsoft VS Code is recommended.
- Sections 10.4–10.6: A .NET MAUI IDE is required. Microsoft Visual Studio Community edition is recommended.

Figure 10.5. The provided web map has a feature layer of world countries with generalized geometries.

10.1: Create a responsive web app using HTML and CSS

This section requires extensive typing. To save time, you can copy the necessary code snippets directly from the key, located at chapter10\step1_key.html. This can also serve as a reference for correctly inserting code snippets.

1. If you haven't already done so, download the provided tutorial source code from https://arcg.is/1Orn9a1 and unzip it.

2. On your computer, locate the files unzipped, open chapter10\step1.html in a web browser, and review its functions.

   The app displays a web map of world countries.

3. In a JavaScript editor, open chapter10\step1.html.

4. Review the source code.

   The source code is based on the "Load a Basic WebMap" sample of Maps SDK for JavaScript. You can see the original sample and explanation at arcg.is/2fd7ZYK.

   Next, you will add a new quiz div for displaying questions and showing whether users' answers are correct or wrong. The <div> tag in HTML defines a division or section, serving as a container for other HTML elements.

5. Find the existing line in the source code:
   ```
   <div id="viewDiv"></div>
   ```

6. Under the existing line, add the subsequent snippet. Reference the key as needed.
   ```
   <div id="quizDiv">
   </div>
   ```

   Next, you will use CSS media queries to style the layout for portrait mode. Position the map view div to occupy the top 75% of the browser window and place the quiz div in the bottom 25%.

   Extra lines of space are not required but are added to make the code more legible.

7. Find the CSS section, indicated by the style tags, located between lines 8 and 18.

8. Add the following snippet before the closing style tag. Reference the key as needed.

```
@media (orientation: portrait) {
#viewDiv {
position: absolute;
top: 0px;
left: 0px;
width: 100%;
height: 50%;
}

#quizDiv {
position: absolute;
bottom: 0px;
left: 0px;
width: 100%;
height: 50%;
}
}
```

Next, you will use a CSS media query to style the layout for landscape mode. Place the map view div on the left 60% of the browser window and display the quiz div on the right 40%.

9. Add the following to the end of the last snippet:
   ```
   @media (orientation: landscape) {
   #viewDiv {
   position: absolute;
   top: 0px;
   left: 0px;
   width: 60%;
   height: 100%;
   }

   #quizDiv {
   position: absolute;
   top: 0px;
   right: 0px;
   width: 40%;
   height: 100%;
   }
   }
   ```

10. Remove the comma after body and remove #viewDiv from the CSS as shown with a strike-through.
    ```
    html,
    body~~,~~
    ~~#viewDiv~~ {
    padding: 0;
    margin: 0;
    height: 100%;
    width: 100%;
    }
    ```

    This action will ensure the view div styling relies on the definitions you provided in the previous two steps.

11. Save your source code.

12. In your web browser, refresh the file chapter10\step1.html and resize your browser to switch between landscape and portrait modes. Observe the layout changes accordingly.

    To simulate specific mobile devices, you can switch your web browser to mobile mode.

13. Open your web browser's developer tools by clicking the Customize and Control Google Chrome button (three dots in the upper right of the page). Click More Tools and choose Developer Tools.

14. Click the Toggle Device toolbar.

15. Select a mobile device from the Dimensions list, and then click the Rotate button to switch between portrait and landscape orientations. Observe the layout changes as previously defined.

10.2: Generate quiz questions using JavaScript

In this section, you will begin by querying the countries layer to compile a list of countries. Then, you will randomly select one country and challenge users to locate it on the map.

This section requires extensive typing. To save time, you can copy the necessary code snippets directly from the key, located at chapter10\step2_key.html. This can also serve as a reference for correctly inserting code snippets.

1. In your JavaScript editor, open chapter10\step2.html and review its functions.

   Step2.html contains the source code you obtained at the end of the previous section.

2. Find the HTML body, indicated by the body tags, located between lines 74 and 78.

3. In the HTML body, insert the following snippet after the quiz div to display the quiz questions. Your code should look like this. Reference the key as needed.
   ```
   <div id="quizDiv">
     <div id="questionDiv"></div>
   </div>
   ```

4. Find the JavaScript section, indicated by the script tags, located between lines 59 and 72.

5. In the JavaScript section, add the last two lines in the code snippet to the source code as illustrated.

```
container: "viewDiv"
});
let countries = [];
let countryInQuestion = null;
```

These lines declare two variables: one to store the list of countries and another for the country currently featured in the quiz.

6. Below the two snippet lines you just added, insert the following function. You may skip the comment lines or copy the source code directly from the key for convenience.

```
function queryTheCountries() {
// Get the countries layer
countriesLayer = webmap.layers.getItemAt(0);
// Create a query param variable to store the parameters
let queryparams = countriesLayer.createQuery();
queryparams.returnGeometry = true;
queryparams.where = "COUNTRY is not null";
// Order the countries by their areas descending
queryparams.orderByFields = ["Shape__Area DESC"];
// Return the first 20 countries
queryparams.num = 20;
// Execute the query
countriesLayer.queryFeatures(queryparams).then(function (results) {
// Store the countries in the countries array
countries = results.features;
showQuestion();
});
}
```

This function queries the countries layer to return the 20 largest countries. After the query is complete, it saves the results for future use and calls a function named showQuestion, which you will implement in the next step.

7. Below the function you just added, insert the following function.

```
function showQuestion() {
// Get the number of countries
let numCountries = countries.length;
// Generate a random number between 0 and the number of
countries
let randomNum = Math.floor(Math.random() * numCountries);
// Get the country at the random position
countryInQuestion = countries[randomNum];
// Get the country name
let countryName = countries[randomNum].attributes["COUNTRY"];
// Get the question div
let questionDiv = document.getElementById("questionDiv")
// Display the question in the question div
questionDiv.innerHTML = "On the map tap " + countryName;
}
```

This new function selects a random country and displays the question in the question div. The comments within the function offer detailed explanations.

8. Below the function you just added, insert the following code snippet:
```
view.when(() => {
// Query the countries
queryCountries();
});
```

These lines invoke the queryCountries function once the map view is initialized.

9. Save your source code.

10. Refresh chapter10\step2.html in your web browser to see a quiz question prompting users to locate and tap on a specified country.

> Note: You can modify the game's difficulty level by adjusting the number of countries included in the queryCountries function. Using the 20 largest countries makes the game relatively easy because these are easier to find. Increasing the number to include smaller countries makes the game more challenging because these countries are typically harder to locate.

10.3: Monitor user tapping and evaluate user answers using JavaScript

This section will add a listener for tap or click events on the map view to query the country name and check whether it matches the one in the quiz.

This section requires extensive typing. To save time, you can copy the necessary code snippets directly from the key, located at chapter10\step3_key.html. This can also serve as a reference for correctly inserting code snippets.

1. In your JavaScript editor, open chapter10\step3.html.

   Step3.html has the source code from the end of the previous section.

2. In the HTML body, add the following snippet under the question div to display the answers. Refer to the key as needed.

   ```
   <div id="quizDiv">
    <div id="questionDiv"></div>
    <div id="answerDiv"></div>
   </div>
   ```

   Next, you will add an event listener to monitor map click or tap events.

3. In the JavaScript section, add the following snippet below the line to query countries:

   ```
   view.when(() => {
     // Query the countries
     queryCountries();
     // Listen to the click event on the view
     view.on("click", handleMapClick);
   });
   ```

   The code will trigger a function named `handleMapClick` whenever users click or tap on the map.

   Next, you will implement the `handleMapClick` function.

4. Below the `showQuestion` function, add the following function.

   ```
   function handleMapClick(event) {
     // get the map point from the click event
     let mapPoint = event.mapPoint;
   ```

```
// Create a query to get the country at the clicked location
let queryparamsClick = countriesLayer.createQuery();
queryparamsClick.geometry = mapPoint;
queryparamsClick.spatialRelationship = "intersects";
queryparamsClick.returnGeometry = false;
queryparamsClick.outFields = ["COUNTRY"]
// Execute the query
countriesLayer.queryFeatures(queryparamsClick).then(function
(results) {
checkAnswer(results);
});
}
```

This function performs a query on the country layer to retrieve the name of the country at the location where the user clicked or tapped. When the query returns, the results are passed to a function named checkAnswer, which you will add in the next step.

5. Below the `handleMapClick` function, add the following function.

```
function checkAnswer(results) {
  // If no countries were returned, display a message in the answer div
  if (results.features.length == 0) {
  document.getElementById("answerDiv").innerHTML = "No countries were tapped";
        return;
```

```
    }
    // Get the clicked country
    let countryClicked = results.features[0];
    // Get the name of the clicked country
    const countryNameClicked = countryClicked.
attributes["COUNTRY"];
    // Check if the clicked country is the correct country
    let msg = ""
    if (countryNameClicked == countryInQuestion.attributes["COUN-
TRY"]) {
    msg = "Correct!";
    } else {
    msg = "Wrong! That was " + countryNameClicked;
    }
    // Display the result message in the answer div
    document.getElementById("answerDiv").innerHTML = msg;
    }
```

This function checks if the country tapped by the user is the correct one.

6. Save your source code.

7. Refresh chapter10\step3.html in your web browser, read the displayed question, and tap a country on the map. Review the result message to see whether you tapped the correct country.

Next, you will add code to generate a new question two seconds after evaluating the user's answer.

8. Add the lines highlighted below.
```
    // Display the result message in the answer div
    document.getElementById("answerDiv").innerHTML = msg;
    // Wait for 2 seconds, then display a new question
    setTimeout(function () {
    showQuestion();
    }, 2000);
```

9. Add the highlighted lines to the `showQuestion` function to clear the answer div after displaying a new question.

```
// Display the question in the question div
questionDiv.innerHTML = "On the map tap " + countryName;
// Clear the answer div
document.getElementById("answerDiv").innerHTML = "";
}
```

10. Save your source code.

11. Refresh chapter10\step3.html in your web browser.

    A new question is presented after you answer the previous one.

    In this section, you developed a responsive web app and tested it in the mobile views of your web browsers. For a real project, you would need to deploy it to a web server and test it on phones and tablets before releasing the app.

10.4: Create a responsive native app using .NET MAUI

In this and the next two sections, you will develop a native version of the Country Tutor game.

Prerequisites: Before you continue, ensure that you have done the following:
- Installed Microsoft Visual Studio.
- Installed the ArcGIS Maps SDK for .NET, the ArcGIS Maps SDK for .NET Toolkit, and Visual Studio project templates. Refer to the "Install and Set Up" section at https://arcg.is/1Orn9a1.
- Obtained an ArcGIS Developer API Key with Location services privilege enabled. This requires an ArcGIS Online Administrator account. Refer to the "Create an API Key" section at https://arcg.is/1Orn9a1.
- Downloaded and unzipped the tutorial source code provided at https://arcg.is/1Orn9a1.

This section requires extensive typing. To save time, you can copy the necessary code snippets directly from the key files, located at chapter10\Section_10_4, including Styles.axml, MainPage.xaml, and MainPage.xaml.cs.

1. Start Visual Studio.

2. In the Visual Studio start screen, click Create a New Project.

3. Choose the ArcGIS Maps SDK .NET MAUI App (Esri) template. If you don't see the template listed, you can find it by typing MAUI in the Search for Templates text box.

4. Click Next.

5. For Project name, type CountryTutor. For Location, save to your Chapter 10 folder.

6. Click Next.

7. For Framework, choose .NET 8.

8. For Create a 2D map or 3D scene, choose Map.

9. Click Create to create the project.

   Next, you will add a reference to ArcGIS Maps SDK for .NET MAUI.

10. Under Solution Explorer, right-click Dependencies and click Manage NuGet Packages.

11. In the upper right of the NuGet Package Manager window, ensure that the selected Package source is nuget.org.

12. Click the Browse tab and search for Esri.ArcGISRuntime.Maui.

13. In the search results, click the Esri.ArcGISRuntime.Maui NuGet package.

14. Confirm that the Latest Stable Version of the package is selected in the Version list.

15. Click Install.

    The Preview Changes dialog box confirms any package dependencies or conflicts.

16. Review the changes and click Apply to continue installing the packages.

17. Review the license information on the License Acceptance dialog box and click I Accept to add the package(s) to your project.

18. In the Visual Studio Output window, ensure that the packages were successfully installed.

19. Close the NuGet Package Manager window.

20. In Visual Studio Solution Explorer, click MauiProgram.cs to open its source code. Uncomment the UseAPIKey line and enter your API Key.
    ```
    .UseArcGISRuntime(config => config
    // .UseLicense("[Your ArcGIS Maps SDK license string]")
    // .UseApiKey("[Your ArcGIS location services API key]")
    .ConfigureAuthentication(auth => auth
    ```

21. Expand MainPage.xaml and click MainPage.xaml.cs to open its source code.

22. At the top of MainPage.xaml.cs, add the following highlighted code to import the required ArcGIS Runtime modules.
    ```
    using Esri.ArcGISRuntime.Portal;
    ```

23. In MainPage.xaml.cs, add the following highlighted code. Refer to the key as needed.

    The code turns myMap into a member variable of the class. The comment lines, introduced by two forward slashes, are provided to help you understand each line of code. You don't need to type the comment lines in the source code.
    ```
    public partial class MainPage : ContentPage
    {
     // Create a new map variable
     private Esri.ArcGISRuntime.Mapping.Map myMap;
     public MainPage()
     {
     InitializeComponent();
     this.BindingContext = new MapViewModel();
     // Call the Initialize function to set up the map
      _ = Initialize();
    ```

```
    }
    private async Task Initialize()
    {
    // Create a portal. If a URI is not specified, www.arcgis.com is
    used by default.
    ArcGISPortal portal = await ArcGISPortal.CreateAsync();
    // Get the portal item for a web map using its unique item id.
    PortalItem mapItem = await PortalItem.CreateAsync(portal,
    "faa4aefba8a14fd58253dbf838f35a82");
    // Create the map from the item.
    myMap = new Esri.ArcGISRuntime.Mapping.Map(mapItem);
    // Create a map with an initial viewpoint.
    mapView.Map = myMap;
    // wait for the map to load
    await myMap.LoadAsync();
    }
}
```

24. On the Visual Studio title bar, click Build and choose Build Solution.

25. From the device list, click Android Emulators and choose an Android emulator.

    If you don't have an Android emulator, you can create one using ADM. You can open ADM by clicking Tools > Android > Android Device Manager, or by clicking the ADM button (phone with an arrow) on the Visual Studio toolbar.

26. On the toolbar, click the Run button.

    It may take a few minutes for the emulator to start. Once the code runs, you will see the World Countries web map displayed. Navigating the map in the emulator differs from using an actual multitouch mobile screen.

27. Drag the map to pan it. Press and hold Ctrl and then drag your mouse to zoom in and out and rotate the map.

28. On the Visual Studio toolbar, click the Stop button to stop the app.

    Next, you will add the quiz element to display questions and indicate whether users' answers are correct or incorrect. In the code, you will refer to several styles that you will define later.

29. In Visual Studio, click MainPage.xaml to open it and add the following highlighted code. Refer to the key as needed.

```xml
<Grid Style="{StaticResource MainPageStyle}" >
<esri:MapView x:Name="mapView" Style="{StaticResource MapViewStyle}"/>
<StackLayout x:Name="quizView" Style="{StaticResource QuizViewStyle}">
<Label x:Name="questionLbl" Style="{StaticResource LabelStyle}"/>
<Label x:Name="answerLbl" Style="{StaticResource LabelStyle}"/>
</StackLayout>
</Grid>
```

    Next, you will define the styles referenced in the previous step by editing Resources\Styles\Styles.xaml.

30. Replace the existing code in Resources\Styles\Styles.xaml with the following code. Refer to the key as needed.

```xml
<?xml version="1.0" encoding="UTF-8" ?>
<?xaml-comp compile="true" ?>
<ResourceDictionary xmlns="http://schemas.microsoft.com/dotnet/2021/maui"
  xmlns:x="http://schemas.microsoft.com/winfx/2009/xaml"
  xmlns:esri="http://schemas.esri.com/arcgis/runtime/2013">
  <Style x:Key="MainPageStyle" TargetType="Grid">
  <Setter Property="VisualStateManager.VisualStateGroups">
  <VisualStateGroupList>
  <VisualStateGroup>
  <VisualState x:Name="Vertical">
  <VisualState.StateTriggers>
```

```xml
<OrientationStateTrigger Orientation="Portrait" />
</VisualState.StateTriggers>
<VisualState.Setters>
<Setter Property="RowDefinitions" Value="2*,*" />
</VisualState.Setters>
</VisualState>
<VisualState x:Name="Horizontal">
<VisualState.StateTriggers>
<OrientationStateTrigger Orientation="Landscape" />
</VisualState.StateTriggers>
<VisualState.Setters>
<Setter Property="ColumnDefinitions" Value="2*,*" />
</VisualState.Setters>
</VisualState>
</VisualStateGroup>
</VisualStateGroupList>
</Setter>
</Style>
<Style x:Key="MapViewStyle" TargetType="{x:Type esri:MapView}">
<Setter Property="Grid.Row" Value="0" />
<Setter Property="Grid.Column" Value="0" />
</Style>
<Style TargetType="StackLayout" x:Key="QuizViewStyle">
<Setter Property="VisualStateManager.VisualStateGroups">
<VisualStateGroupList>
<VisualStateGroup>
<VisualState x:Name="Portrait">
<VisualState.StateTriggers>
<OrientationStateTrigger Orientation="Portrait" />
</VisualState.StateTriggers>
<VisualState.Setters>
<Setter Property="Grid.Row" Value="1" />
<Setter Property="Grid.Column" Value="0" />
</VisualState.Setters>
</VisualState>
<VisualState x:Name="Landscape">
<VisualState.StateTriggers>
<OrientationStateTrigger Orientation="Landscape" />
</VisualState.StateTriggers>
```

```
            <VisualState.Setters>
            <Setter Property="Grid.Row" Value="0" />
            <Setter Property="Grid.Column" Value="1" />
            </VisualState.Setters>
            </VisualState>
            </VisualStateGroup>
            </VisualStateGroupList>
            </Setter>
        </Style>
        <Style TargetType="Label" x:Key="LabelStyle">
            <Setter Property="FontAttributes" Value="Bold" />
            <Setter Property="Margin" Value="10" />
        </Style>
    </ResourceDictionary>
```

The code defines the following styles:
- **MainPageStyle:** Makes the app responsive. For landscape orientation, it divides the screen into one row and two columns, with the left column taking up 2/3 of the width and the right one 1/3.
- **MapViewStyle:** Sets the map view to occupy the grid at row 0 and column 0.
- **QuizViewStyle:** Positions the quiz view element in the grid at row 0 and column 1 in landscape mode and at row 1 and column 0 otherwise.
- **LabelStyle:** Styles the labels to be bold and to have a margin of 10 pixels on all sides.

31. Save and rebuild your project.

32. Run your code in the Android emulator or another platform you have configured.

33. Once the app is running in the emulator, click Rotate on the Emulator toolbar and observe how the app responds to the orientation changes.

10.5: Generate quiz questions using C#

This section will implement logic like that of section 10.2. It will first query the countries layer to obtain a list of countries, randomly select one, and ask users to find it on the map. You will primarily work with MainPage.xaml.cs.

This section requires extensive typing. To save time, you can copy the necessary code snippets directly from the key files, located at chapter10\Section_10_5.

1. Start Visual Studio and open the project you created in the previous section.

2. In MainPage.xaml.cs, add the following two highlighted lines at the top to import all the needed types from the namespaces.
   ```
   using Esri.ArcGISRuntime.Data;
   using Esri.ArcGISRuntime.Mapping;
   using Esri.ArcGISRuntime.Portal;
   ```

3. In MainPage.xaml.cs, add the two lines highlighted below.

   These lines declare two variables to store the countries and the country featured in the quiz.
   ```
   public partial class MainPage : ContentPage
   {
   private Feature countryInQuestion = null;
   private Feature[] countries = null;
   private Esri.ArcGISRuntime.Mapping.Map myMap;
   ```

4. Below the Initialize function, add the following function. Refer to the key as needed.
   ```
   private async void queryTheCountries()
   {
    // Get the countries layer
    FeatureLayer countriesLayer = (FeatureLayer)myMap.OperationalLayers[0];
    // Create a query param variable to store the parameters
    QueryParameters queryCountries = new QueryParameters
    {
    ReturnGeometry = true,
    WhereClause = "COUNTRY is not null"
    };
    // Order the countries by their areas descending
    queryCountries.OrderByFields.Add(new OrderBy("Shape__Area", SortOrder.Descending));
    //queryCountries.OrderByFields.Add(new OrderBy("COUNTRY", SortOrder.Ascending));
    // Return the first 20 countries
    queryCountries.MaxFeatures = 20;
   ```

```
    // Execute the query
    FeatureQueryResult resultCountries = await countriesLayer.
FeatureTable.QueryFeaturesAsync(queryCountries);
    // Store the countries in the countries array
    this.countries = resultCountries.ToArray();
    showQuestion();
}
```

This function queries the countries layer to retrieve the 20 largest countries. After the query is completed, it stores the results for future use and calls the showQuestion function, which you will define later.

5. Below the function you just added, add the following function.

This function selects a random country and displays the question in the question label.
```
    private void showQuestion()
    {
      // Get the number of countries
      int numCountries = this.countries.Length;
      // Generate a random number between 0 and the number of
    countries
      Random random = new Random();
      int randomIndex = random.Next(0, numCountries);
      // Get the country at the random position
      this.countryInQuestion = this.countries[randomIndex];
      // Get the country name
      string countryName = this.countryInQuestion.
    GetAttributeValue("COUNTRY").ToString();
      // Display the question in the question label
      questionLbl.Text = "On the map tap " + countryName;
    }
```

6. In the Initialize function, add the following highlighted lines to call the queryCountries function.
```
        // wait for the map to load
        await myMap.LoadAsync();
        // Query the countries
        queryTheCountries();
    }
```

7. Save your source code.

8. Rebuild your project.

9. Run your project in an emulator.

   It displays a question asking users to tap a country.

   Similar to the JavaScript edition, you can adjust the game's difficulty level by changing the `MaxFeatures` number in the `queryCountries` function.

10.6: Monitor user tapping and evaluate user answers using C#
Similar to section 10.3, this section will add code to query for the country where users tapped and determine whether the answer is correct.

This section requires extensive typing. To save time, you can copy the necessary code snippets directly from the key file, located at chapter10\Section_10_6\MainPage.xaml.cs.

1. Start Visual Studio and continue with the project you completed in the previous section.

   Next, you will add an event listener to monitor map click or tap events.

2. In MainPage.xaml.cs, add the highlighted lines to the Initialize function.
   ```
   // Query the countries
   queryTheCountries();
   ```

   Next, you will add the handleMapClick function.

3. Below the showQuestion function, add the following function.
   ```
   private async void handleMapClick(object sender,
   GeoViewInputEventArgs e)
   {
   // Get the countries layer
   FeatureLayer countriesLayer = (FeatureLayer)myMap.
   OperationalLayers[0];
   // Create a query param variable to store the parameters
   QueryParameters queryCountries = new QueryParameters
   {
   ReturnGeometry = false,
   WhereClause = "COUNTRY is not null",
   ```

```
      Geometry = e.Location,
      SpatialRelationship = SpatialRelationship.Intersects
    };
    // Execute the query
    FeatureQueryResult results = await countriesLayer.FeatureTable.
    QueryFeaturesAsync(queryCountries);
    checkAnswer(results);
}
```

This function queries the countries layer to retrieve the country at the location where the user tapped, and then passes the results to a function named checkAnswer, which will be added in the next step.

4. Under the handleMapClick function, add the following function to check whether the tapped country is correct.

```
    private async void checkAnswer(FeatureQueryResult results)
    {
    // If the query returns no country, display a message in the
    answer label
    if (results.Count() == 0)
    {
    answerLbl.Text = "No countries were tapped";
    return;
    }
    // Get the clicked country
    GeoElement countryClicked = results.ElementAt(0);
    // Get the name of the clicked country
    string countryNameClicked = countryClicked.
    Attributes["COUNTRY"].ToString();
    // Check if the clicked country is the correct country
    string msg = "";
    if (countryNameClicked == this.countryInQuestion.
    GetAttributeValue("COUNTRY").ToString())
    {
    msg = "Correct!";
    }
    else
    {
    msg = "Wrong! That was " + countryNameClicked;
```

```
    }
    // Display the result message in the answer label
    answerLbl.Text = msg;
}
```

5. Save and rebuild your source code.

6. Run your app in an emulator. Read the question it displays and tap a country on the map. Review the result message to confirm that you tapped the correct one.

   Next, you will add code to generate a new question two seconds after evaluating the user's answer.

7. Add the lines highlighted below.
```
    // Display the result message in the answer label
    answerLbl.Text = msg;
    // Wait for 2 seconds, then display a new question
    await Task.Delay(2000);
    showQuestion();
}
```

8. Add the highlighted lines below to the `showQuestion` function. The code will clear the answer label after displaying a new question.
```
    // Display the question in the question label
    questionLbl.Text = "On the map tap " + countryName;
    // Clear the answer label
    answerLbl.Text = "";
}
```

9. Save and build your source code.

10. Run the app in an emulator.

    A new question is presented after you answer the previous question.

    In this tutorial, you learned the basic steps for developing custom mobile apps. You created a Country Tutor game to help users learn about the countries of the world. You developed two editions of the game: one using JavaScript and the other with .NET MAUI. The JavaScript edition is easier to develop, whereas the .NET MAUI edition is more challenging. It stands out for its ability to be compiled into native apps for multiple platforms, such as Android and iOS, offering a cost-efficient way to create cross-platform native apps. Throughout this tutorial, it becomes evident that despite the significant differences between JavaScript and .NET MAUI, the underlying logic that empowers them shares much in common. This similarity underscores a valuable lesson: Mastering a single programming language and one ArcGIS Maps SDK can significantly streamline the process of learning additional languages and SDKs. If your project requires custom solutions, don't feel intimidated to learn a programming language and an ArcGIS Maps SDK or expand your skills to other languages and SDKs as needed.

The apps created in this tutorial are simple yet demonstrate the versatility, potential, and joy of programming. Navigating through programming challenges can be daunting; at times—even a minor typo can lead to hours of debugging. If you experienced frustration during this tutorial, rest assured, you're not alone. It's a sign of true learning. As you deepen your understanding and skills, you'll unlock the full potential of Mobile GIS, enabling you to create captivating and impactful apps that turn your or your organization's creative visions into reality.

Assignment 10: Enhance the tutorial mobile apps' responsive layout and map interaction

Requirements:
- Enhance the JavaScript and .NET MAUI mobile apps developed in this chapter by:
  1. Moving the quiz to the left side in landscape mode and to the top in portrait mode.
  2. Adding a Show Answer button. When users tap it, the app should zoom in on the map to the country being questioned and highlight the country temporarily.

What to submit:
- Screenshots demonstrating how your apps respond to changes in landscape and portrait orientations.
- The source code for your JavaScript/HTML/CSS and the Visual Studio Project files for your .NET MAUI app.

Image and data credits

Chapter 1: All images and data courtesy of Esri.

Chapter 2: None.

Chapter 3: Survey123 images courtesy of Esri.

Chapter 4: None.

Chapter 5: Figures 1–13:
- Esri.
- Esri.
- Esri.
- City of Redlands, County of San Bernardino, Maxar, nearmap.
- City of Redlands, County of San Bernardino, Maxar, nearmap.
- Esri.
- City of Redlands, County of San Bernardino, Maxar, nearmap.
- Esri and Google.
- Esri.
- Esri.
- Esri.
- Esri.
- Esri Community Maps Contributors, Loma Linda University, City of Redlands, County of Riverside, County of San Bernardino, California State Parks, © OpenStreetMap, Microsoft, Esri, TomTom, Garmin, SafeGraph, GeoTechnologies Inc., METI/NASA, USGS, Bureau of Land Management, EPA, NPS, US Census Bureau, USDA, USFWS.

Chapter 6: Figures 1–11:
- Esri, SCE, NASA, NGA, USGS, FEMA | Esri Community Maps Contributors, County of San Bernardino, California State Parks, © OpenStreetMap, Microsoft, Esri, TomTom, Garmin, SafeGraph, GeoTechnologies Inc., METI/NASA, USGS, Bureau of Land Management, EPA, NPS, US Census Bureau, USDA, USFWS.
- Esri.
- Esri.

Esri, HERE, Garmin, SafeGraph, METI/NASA, USGS, EPA, NPS, USDA.

Esri.

Esri.

Esri.

Esri.

Esri.

Esri Community Maps Contributors, County of San Bernardino, California State Parks, © OpenStreetMap, Microsoft, Esri, TomTom, Garmin, SafeGraph, GeoTechnologies Inc., METI/NASA, USGS, Bureau of Land Management, EPA, NPS, US Census Bureau, USDA, USFWS.

Esri Community Maps Contributors, Loma Linda University, City of Redlands, County of Riverside, County of San Bernardino, California State Parks, © OpenStreetMap, Microsoft, Esri, TomTom, Garmin, SafeGraph, GeoTechnologies Inc., METI/NASA, USGS, Bureau of Land Management, EPA, NPS, US Census Bureau, USDA, USFWS.

Chapter 7: Figures 1–18:

Esri.

Esri, NASA, NGA, USGS, FEMA | Esri Community Maps Contributors, Loma Linda University, City of Redlands, County of Riverside, California State Parks, Esri, TomTom, Garmin, SafeGraph, GeoTechnologies Inc., METI/NASA, USGS, Bureau of Land Management, EPA, NPS, US Census Bureau, USDA, USFWS | Oak Ridge National Laboratory (ORNL), National Geospatial-Intelligence Agency (NGA) Homeland Security Infrastructure Program (HSIP) Team.

Esri, NASA, NGA, USGS, FEMA | Esri Community Maps Contributors, Loma Linda University, City of Redlands, County of Riverside, California State Parks, Esri, TomTom, Garmin, SafeGraph, GeoTechnologies Inc., METI/NASA, USGS, Bureau of Land Management, EPA, NPS, US Census Bureau, USDA, USFWS | Oak Ridge National Laboratory (ORNL), National Geospatial-Intelligence Agency (NGA) Homeland Security Infrastructure Program (HSIP) Team.

Esri, NASA, NGA, USGS, FEMA | Esri Community Maps Contributors, County of San Bernardino, California State Parks, © OpenStreetMap, Microsoft, Esri, TomTom, Garmin, SafeGraph, GeoTechnologies Inc., METI/NASA, USGS, Bureau of Land Management, EPA, NPS, US Census Bureau, USDA, USFWS.

Esri.

Esri.

Esri.

Esri.

Esri and Google.

Esri.

Image and data credits **265**

Esri.
Esri.
Esri.
Esri.
Esri.
Esri.
Esri.
Esri and Google.

Chapter 8: Figures 1–17:

Esri.
Esri.
Loma Linda University, City of Redlands, County of Riverside, County of San Bernardino, California State Parks, © OpenStreetMap, Microsoft, Esri, TomTom, Garmin, SafeGraph, GeoTechnologies Inc., METI/NASA, USGS, Bureau of Land Management, EPA, NPS, US Census Bureau, USDA, USFWS, NGA, and FEMA.
Esri and Microsoft.
Esri and Microsoft.
Esri and Microsoft.
Esri and Microsoft.
Esri.
Esri.
Esri and Microsoft.
Esri and Microsoft.
Esri and Microsoft.
Esri and Microsoft.
Esri and Microsoft.
Esri and Microsoft.
Esri and Microsoft.
Esri and Microsoft.

Chapter 9: Figures 1–20:

Esri.
Esri.
Esri.
Esri and TensorFlow.
Esri.
Esri.

County of San Bernardino, Maxar, Microsoft | Source: Airbus, USGS, NGA, NASA, CGIAR, NLS, OS, NMA, Geodatastyrelsen, GSA, GSI, and the GIS User Community | Esri Community Maps Contributors, City of Rancho Cucamonga, County of San Bernardino, California State Parks, © OpenStreetMap, Microsoft, Esri, TomTom, Garmin, SafeGraph, GeoTechnologies, Inc, METI/NASA, USGS, Bureau of Land Management, EPA, NPS, US Census Bureau, USDA, USFWS

County of San Bernardino, Maxar, Microsoft | Source: Airbus, USGS, NGA, NASA, CGIAR, NLS, OS, NMA, Geodatastyrelsen, GSA, GSI, and the GIS User Community | Esri Community Maps Contributors, City of Rancho Cucamonga, County of San Bernardino, California State Parks, © OpenStreetMap, Microsoft, Esri, TomTom, Garmin, SafeGraph, GeoTechnologies Inc., METI/NASA, USGS, Bureau of Land Management, EPA, NPS, US Census Bureau, USDA, USFWS.

Esri.

Esri.

Esri.

Esri.

Esri.

Esri.

Esri.

Esri and TensorFlow.

Esri Community Maps Contributors, Loma Linda University, City of Redlands, County of Riverside, County of San Bernardino, California State Parks, © OpenStreetMap, Microsoft, Esri, TomTom, Garmin, SafeGraph, GeoTechnologies Inc., METI/NASA, USGS, Bureau of Land Management, EPA, NPS, US Census Bureau, USDA, USFWS.

Esri.

Esri.

Esri.

Chapter 10: Two figures and data courtesy of Esri.

About Esri Press

Esri Press is an American book publisher and part of Esri, the global leader in geographic information system (GIS) software, location intelligence, and mapping. Since 1969, Esri has supported customers with geographic science and geospatial analytics, what we call The Science of Where. We take a geographic approach to problem-solving, brought to life by modern GIS technology, and are committed to using science and technology to build a sustainable world.

At Esri Press, our mission is to inform, inspire, and teach professionals, students, educators, and the public about GIS by developing print and digital publications. Our goal is to increase the adoption of ArcGIS and to support the vision and brand of Esri. We strive to be the leader in publishing great GIS books, and we are dedicated to improving the work and lives of our global community of users, authors, and colleagues.

Acquisitions
Stacy Krieg
Claudia Naber
Alycia Tornetta
Jenefer Shute

Product Engineering
Craig Carpenter
Maryam Mafuri

Editorial
Carolyn Schatz
Mark Henry
David Oberman

Production
Monica McGregor
Victoria Roberts

Sales & Marketing
Eric Kettunen
Sasha Gallardo
Beth Bauler

Contributors
Christian Harder
Matt Artz

Business
Catherine Ortiz
Jon Carter
Jason Childs

Related Titles

Designing Map Interfaces
Michael Gaigg
9781589487253

Top 20 Essential Skills for ArcGIS Online
Craig Carpenter et al.
9781589487802

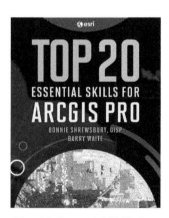

Top 20 Essential Skills for ArcGIS Pro
Bonnie Shrewsbury
Barry Waite
9781589487505

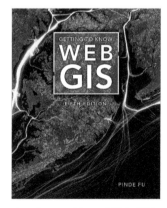

Getting to Know Web GIS, fifth edition
Pinde Fu
9781589487277

For information on Esri Press books, e-books, and resources, visit our website at
esripress.com